THE NATIONAL ACADEMIES
KECK FUTURES INITIATIVE

COMPLEX SYSTEMS

TASK GROUP SUMMARIES

Conference
Arnold and Mabel Beckman Center
Irvine, California
November 13 —15, 2008

THE NATIONAL ACADEMIES PRESS
Washington, D.C.
www.nap.edu

THE NATIONAL ACADEMIES PRESS   500 Fifth Street, N.W.   Washington, DC 20001

NOTICE: The task group summaries in this publication are based on task group discussions during the National Academies Keck *Futures Initiative* Conference on Complex Systems held at the Arnold and Mabel Beckman Center in Irvine, California, November 13-15, 2008. The discussions in these groups were summarized by the authors and reviewed by the members of each task group. Any opinions, findings, conclusions, or recommendations expressed in this publication are those of the task groups and do not necessarily reflect the view of the organizations or agencies that provided support for this project. For more information on the National Academies Keck *Futures Initiative* visit www.keckfutures.org.

Funding for the activity that led to this publication was provided by the W.M. Keck Foundation. Based in Los Angeles, the W. M. Keck Foundation was established in 1954 by the late W.M. Keck, founder of the Superior Oil Company. The Foundation's grant making is focused primarily on pioneering efforts in the areas of science and engineering research; undergraduate education; medical research; and Southern California. Each grant program invests in people and programs that are making a difference in the quality of life, now and for the future. For more information visit www.wmkeck.org.

International Standard Book Number-13:   978-0-309-13725-6
International Standard Book Number-10:   0-309-13725-X

Additional copies of this report are available from the National Academies Press, 500 Fifth Street, N.W., Lockbox 285, Washington, DC 20055; (800) 624-6242 or (202) 334-3313 (in the Washington metropolitan area); Internet, http://www.nap.edu.

Copyright 2009 by the National Academy of Sciences. All rights reserved.

Printed in the United States of America

THE NATIONAL ACADEMIES
Advisers to the Nation on Science, Engineering, and Medicine

The **National Academy of Sciences** is a private, nonprofit, self-perpetuating society of distinguished scholars engaged in scientific and engineering research, dedicated to the furtherance of science and technology and to their use for the general welfare. Upon the authority of the charter granted to it by the Congress in 1863, the Academy has a mandate that requires it to advise the federal government on scientific and technical matters. Dr. Ralph J. Cicerone is president of the National Academy of Sciences.

The **National Academy of Engineering** was established in 1964, under the charter of the National Academy of Sciences, as a parallel organization of outstanding engineers. It is autonomous in its administration and in the selection of its members, sharing with the National Academy of Sciences the responsibility for advising the federal government. The National Academy of Engineering also sponsors engineering programs aimed at meeting national needs, encourages education and research, and recognizes the superior achievements of engineers. Dr. Charles M. Vest is president of the National Academy of Engineering.

The **Institute of Medicine** was established in 1970 by the National Academy of Sciences to secure the services of eminent members of appropriate professions in the examination of policy matters pertaining to the health of the public. The Institute acts under the responsibility given to the National Academy of Sciences by its congressional charter to be an adviser to the federal government and, upon its own initiative, to identify issues of medical care, research, and education. Dr. Harvey V. Fineberg is president of the Institute of Medicine.

The **National Research Council** was organized by the National Academy of Sciences in 1916 to associate the broad community of science and technology with the Academy's purposes of furthering knowledge and advising the federal government. Functioning in accordance with general policies determined by the Academy, the Council has become the principal operating agency of both the National Academy of Sciences and the National Academy of Engineering in providing services to the government, the public, and the scientific and engineering communities. The Council is administered jointly by both Academies and the Institute of Medicine. Dr. Ralph J. Cicerone and Dr. Charles Vest are chair and vice chair, respectively, of the National Research Council.

www.national-academies.org

THE NATIONAL ACADEMIES KECK *FUTURES INITIATIVE* COMPLEX SYSTEMS STEERING COMMITTEE

H. EUGENE STANLEY (Chair) (NAS), University Professor, Professor of Physics, Professor of Biomedical Engineering, Professor of Physiology (School of Medicine) and Director, Center for Polymer Studies, Boston University

ALBERT-LÁSZLÓ BARABÁSI, Distinguished University Professor, Center for Complex Network Research, Department of Physics, Northeastern University; Department of Medicine, Harvard University

JAMES B. BASSINGTHWAIGHTE (NAE), Professor of Bioengineering and Radiology, University of Washington

BONNIE L. BASSLER (NAS), Investigator, Howard Hughes Medical Institute, Professor of Molecular Biology, Department of Molecular Biology, Princeton University

DAVID K. CAMPBELL, Professor of Physics and Electrical Engineering and Provost, Boston University

SALLIE W. CHISHOLM (NAS), Lee and Geraldine Martin Professor of Environmental Studies, Department of Civil and Environmental Engineering, Department of Biology, Massachusetts Institute of Technology

JAMES S. LANGER (NAS), Professor, Department of Physics, University of California, Santa Barbara

SIMON A. LEVIN (NAS), George M. Moffett Professor of Biology, Department of Ecology and Evolutionary Biology, Princeton University

M. ELISABETH PATÉ-CORNELL (NAE), Burt and DeeDee McMurtry Professor and Chair, Department of Management Science and Engineering, Stanford University

MICHAEL A. SAVAGEAU (IOM), Distinguished Professor, Department of Biomedical Engineering and Microbiology Graduate Group, The University of California, Davis

DAVID VALLE (IOM), Henry J. Knott Professor and Director, McKusick-Nathans Institute of Genetic Medicine, Johns Hopkins University

MARC VIDAL, Director, Center for Cancer Systems Biology and Department of Cancer Biology, Harvard Medical School

Staff

KENNETH R. FULTON, Executive Director
KIMBERLY A. SUDA-BLAKE, Program Director
ANNE HEBERGER, Senior Evaluation Associate
RACHEL LESINSKI, Program Associate

The National Academies Keck *Futures Initiative*

THE NATIONAL ACADEMIES KECK *FUTURES INITIATIVE*

The National Academies Keck *Futures Initiative* was launched in 2003 to stimulate new modes of scientific inquiry and break down the conceptual and institutional barriers to interdisciplinary research. The National Academies and the W. M. Keck Foundation believe that considerable scientific progress will be achieved by providing a counterbalance to the tendency to isolate research within academic fields. The *Futures Initiative* is designed to enable scientists from different disciplines to focus on new questions, upon which they can base entirely new research, and to encourage and reward outstanding communication between scientists as well as between the scientific enterprise and the public.

The *Futures Initiative* includes three main components:

Futures Conferences

The *Futures* Conferences bring together some of the nation's best and brightest researchers from academic, industrial, and government laboratories to explore and discover interdisciplinary connections in important areas of cutting-edge research. Each year, some 150 outstanding researchers are invited to discuss ideas related to a single cross-disciplinary theme. Participants gain not only a wider perspective but also, in many instances, new insights and techniques that might be applied in their own work. Additional pre- or post-conference meetings build on each theme to foster further communication of ideas.

Selection of each year's theme is based on assessments of where the intersection of science, engineering, and medical research has the greatest potential to spark discovery. The first conference explored *Signals, Decisions, and Meaning in Biology, Chemistry, Physics, and Engineering*. The 2004 conference focused on *Designing Nanostructures at the Interface between Biomedical and Physical Systems*. The theme of the 2005 conference was *The Genomic Revolution: Implications for Treatment and Control of Infectious Disease*. In 2006 the conference focused on *Smart Prosthetics: Exploring Assistive Devices for the Body and Mind*. In 2007 the conference explored *The Future of Human Healthspan: Demography, Evolution, Medicine and Bioengineering*. In 2008 the conference focused on *Complex Systems* and the 2009 conference will explore *Synthetic Biology*.

Futures Grants

The *Futures* Grants provide seed funding to *Futures* Conference participants, on a competitive basis, to enable them to pursue important new ideas and connections stimulated by the conferences. These grants fill a critical missing link between bold new ideas and major federal funding programs, which do not currently offer seed grants in new areas that are considered risky or exotic. These grants enable researchers to start developing a line of inquiry by supporting the recruitment of students and postdoctoral fellows, the purchase of equipment, and the acquisition of preliminary data—which in turn can position the researchers to compete for larger awards from other public and private sources.

National Academies Communication Awards

The Communication Awards are designed to recognize, promote, and encourage effective communication of science, engineering, medicine, and interdisciplinary work within and beyond the scientific community. Each year the *Futures Initiative* awards $20,000 prizes to those who have advanced the public's understanding and appreciation of science, engineering, and/or medicine. The awards are given in four categories: books, newspaper/magazine, online/Internet, and TV/radio/film. The winners are honored during *Futures* Conferences.

NAKFI cultivates science writers of the future by inviting graduate students from six science writing programs across the country to attend the conference and develop task group discussion summaries and a conference overview for publication in this book. Students are selected by the department director or designee, and prepare for the conference by reviewing the Webcast tutorials and suggested reading, and selecting a task group in which they would like to participate. Students then work with NAKFI's science writing student mentor to finalize their reports following the conferences.

Facilitating Interdisciplinary Research Study

During the first 18 months of the Keck *Futures Initiative*, the Academies undertook a study on facilitating interdisciplinary research. The study examined the current scope of interdisciplinary efforts and provided recommendations as to how such research can be facilitated by funding organizations and academic institutions. *Facilitating Interdisciplinary Research* (2005) is available from the National Academies Press (www.nap.edu) in print and free PDF versions.

About the National Academies

The National Academies comprise the National Academy of Sciences, the National Academy of Engineering, the Institute of Medicine, and the National Research Council, which perform an unparalleled public service by bringing together experts in all areas of science and technology, who serve as volunteers to address critical national issues and offer unbiased advice to the federal government and the public. For more information, visit www.nationalacademies.org.

About the W. M. Keck Foundation

Based in Los Angeles, the W.M. Keck Foundation was established in 1954 by the late W.M. Keck, founder of the Superior Oil Company. The Foundation's grant making is focused primarily on pioneering efforts in the areas of science and engineering; undergraduate education; medical research; and Southern California. Each grant program invests in people and programs that are making a difference in the quality of life, now and in the future. For more information visit www.wmkeck.org.

National Academies Keck *Futures Initiative*
100 Academy
Irvine, CA 92617
949-721-2270 (Phone)
949-721-2216 (Fax)
www.keckfutures.org

Preface

At the National Academies Keck *Futures Initiative* Conference on Complex Systems, participants were divided into twelve interdisciplinary working groups. The groups spent nine hours over two days exploring diverse challenges at the interface of science, engineering, and medicine. The composition of the groups was intentionally diverse, to encourage the generation of new approaches by combining a range of different types of contributions. The groups included researchers from science, engineering, and medicine, as well as representatives from private and public funding agencies, universities, businesses, journals, and the science media. Researchers represented a wide range of experience—from postdoc to those well established in their careers—from a variety of disciplines that included science and engineering, medicine, physics, biology, math/computer science, behavioral science and economics/finance.

The groups needed to address the challenge of communicating and working together from a diversity of expertise and perspectives as they attempted to solve complicated, interdisciplinary problems in a relatively short time. Each group decided on its own structure and approach to tackle the problem. Some groups decided to refine or redefine their problems based on their experience.

Each group presented two brief reports to all participants: (1) an interim report on Friday to debrief on how things were going, along with any special requests; and (2) a final briefing on Saturday, when each group:

- Provided a concise statement of the problem.
- Outlined a structure for its solution.
- Identified the most important gaps in science and technology and recommended research areas needed to attack the problem.
- Indicated the benefits to society if the problem could be solved.

Each task group included a graduate student in a university science writing program. Based on the group interaction and the final briefings, the students wrote the following summaries, which were reviewed by the group members. These summaries describe the problem and outline the approach taken, including what research needs to be done to understand the fundamental science behind the challenge, the proposed plan for engineering the application, the reasoning that went into it and the benefits to society of the problem solution. Due to the popularity of some topics, two groups were assigned to explore the subjects.

Nine webcast tutorials were held in September to help bridge the gaps in terminology used by the various disciplines. Participants had the opportunity to ask questions of the webcast speakers during the live broadcast in September and the panel discussion, which took place immediately prior to the task group breakout sessions.

Contents

Conference Summary 1

TASK GROUP SUMMARIES

1. How would you design the acquisition and organization of the data required to completely model human biology? 5

2. What does it take to achieve a sustainable future? The problem of the commons: achieving a sustainable quality of life.

3. How can we enhance the robustness via interconnectivity? ... 19
   Task Group Summary, Group A, 21
   Task Group Summary, Group B, 26

4. Can engineering systems and control approaches generate new strategies for altering imbalanced macrophage profiles in human disease? 33

5. How can social networks aid our understanding of complexity? ... 41

6. The brain is the epitome of complexity. How will understanding the complex, linked interactions among many types of neurons in the brain lead to knowing how the brain contributes to normal function and susceptibility to neuropsychiatric disease 49
   Task Group Summary, Group A, 51
   Task Group Summary, Group B, 55

7. How can we enhance the robustness of engineered systems, and how can the methods of engineering analysis be extended to address issues of complexity and management in other fields? 59

8. Ecological robustness: Is the biosphere sustainable? 65

9. Can one control flow and transport in complex systems? 73
   Task Group Summary, Group A, 78
   Task Group Summary, Group B, 81

APPENDIXES

Preconference Webcast Tutorials 87
Agenda 91
Participants 95

To view the preconference tutorial webcasts or conference presentations, please visit our website at www.keckfutures.org.

General Summary
Noah Barron

Jorge Luis Borges wrote of an imagined regime that had so mastered cartography that it could create giant, detailed maps, and its crowning achievement was a map of the empire that was the same scale as the empire itself, replicated locations coinciding one-to-one.

> Less attentive to the study of cartography, succeeding generations came to judge a map of such magnitude cumbersome, and, not without irreverence, they abandoned it to the rigors of sun and rain.
>
> "Of Exactitude in Science" from J.L. Borges, *A Universal History of Infamy* (1935)

The challenge presented to the 2008 attendees of the National Academies Keck *Futures Initiative* on Complex Systems was to understand systems as complex as the world itself, with models that are necessarily nuanced to contain enough detail to be useful, while not being so complex as to render those models as impossible to build or understand, and thus relegate them to the deserts of our minds.

Keck board member Richard N. Foster offered a relevant piece of advice to advance thinking on the project; that, like chess masters pondering the small problems of skirmishes and macro problems of the whole game, scientists must learn to effectively "zoom in and zoom out" with their minds.

Network research expert Albert-László Barabási gave the conference a rapid-fire overview of his theories of scale-free networks, that is, naturally emerging but ordered systems whose distributions follow an exponential increase. He described the Internet as "architecture of complexity" that is "driven by our own activities but beyond the comprehension of those who create it."

"How is it possible that they have the same underlying structure," Barabási asked before the panel of hundreds of experts. "And so what? Does it have any consequence?"

Being able to fit a line to the data does not mean we can understand and control the behavior of neurons in a seizing brain or can better build and harden a computer network from attack. But such a grand set of notions helped bring the problems into finer focus for the groups about to meet and tackle their various topics related to complexity. In the words of presenter Dr. Harvey V. Fineberg, president of the Institute of Medicine, the groups decided "which problems are just ripe enough to pick from the tree."

Each task group contained about a dozen experts from many different fields: engineers, microbiologists, computer scientists, economists, mathematicians, paleontologists and neurologists, among others. A graduate science writing student, assigned to every task group, was given the equally stimulating task of somehow capturing the essence of the process, the product and the possibilities that emerged in the discussions.

Task Group 1 tackled the question of how to design the acquisition and organization of the data required to completely model human biology.

Their assignment was to simulate the complex functions of the human body, a job they acknowledged "is larger and more complex than the human body itself."

The group began by drafting a five year plan for collecting and checking biological data, the first step in building the living, breathing simulation of the human body which would not only vastly enhance research but which is an enormous research challenges in and of itself.

Task Group 2 explored what it takes to achieve a sustainable future. The group noted the urgency of the problem, asking, "if not now, when?" and planning a variety of modeling techniques for charting environmental and social degradation, as well as several complex-systems modes for recovery. They concluded that a neural network modeling technique, like that used in brain research, is necessary to tackle nonlinear, holistic issues on planet Earth.

Task Group 3 dealt with the issue of enhancing robustness via interconnectivity. "In many identifiable systems, such as power grid structures, disaster relief networks, airline traffic systems, the Internet and yeast genetic interactions," the group concluded, "the ideal situation that allows robustness to be enhanced is one where both performance optimization and perturbation can be specified." They suggested that vertical connection between hierarchical networks is one solution, as is a "toggle" ability to switch between distributed and centralized organization.

Task Group 4 asked if engineering systems and control approaches can generate new strategies for altering imbalanced macrophage profiles in human disease.

Given the assignment of determining whether cellular and genetic engineering can restrict the growth of cancer, they asked: "Should the focus be on changing M2 macrophages into M1s, or on preventing the development of M2 macrophages in the first place?"

They asked, "Would simply eliminating all M2 macrophages create a tumor fighting phenotype, or should the control system also generate more anti-tumor M1s?"

In the end, Group 4 found that though exerting control on disease spread might be possible, it might be so complex as to be prohibitive. "In the end, it's quite possible that systems to control cell fate might turn out to be just as complex as the organisms they're meant to control."

Task Group 5 pondered social networks' capacity to aid our understanding of complexity. The Internet as a tool for taking snapshots of trends in real life and cyberspace piqued the group's interest. They concluded that mapping a "moving picture" of progressions of diseases and ideologies across the Web would be a valuable first step in using social networks as a already-built sensor system for society.

Task Group 6 handled the brain and the future of understanding the complex, linked interactions among the many types of neurons in the brain. Will that knowledge lead to knowing how the brain contributes to normal function and susceptibility to neuropsychiatric disease?

"An elephant's brain has about four times as many neurons as a human's, but we would assume it is less complex," the group noted.

"It is the organization—not sheer number—of the brain's connections that result in intelligence; complexity captures this organization." And knowing that, how can we engineer computer models, medical interventions or information systems that derive more nuanced function with fewer moving parts?

They settled on an impulse-control model as a metric of complexity, writing "We would compare these different measurements between organisms with different abilities to control their impulses—different strains of mice, different species, humans with certain diseases, and humans with different skills, such as artists and scientists. This will reveal which aspects of the brain's organization are related to impulse control."

Task Group 7 posed questions about enhancing the robustness of engineered systems, and how can the methods of engineering analysis be extended to address issues of complexity and management in other fields. The discussion soon turned to the possibilities of self-regenerating automobiles, space shuttles, a house that could repaint itself, roads that fix their own potholes, and so on. The group's consensus was that by blending the lessons from biology with the lessons from engineering, making machines that heal in a lifelike way could absolutely bolster robustness.

Task Group 8 focused on ecological robustness and in specific, the question of whether the biosphere is sustainable.

Ultimately, the group concluded that preventing the destruction of an ecosystem is far more feasible than rebuilding it. Robustness means not having to clean up the mess in the first place, the group decided. "The models and experiments will reveal projections of ecosystem futures, and what we can do to steer the biosphere towards a path that will sustain humans for generations to come," the group agreed.

Task Group 9 shouldered issues of controlling flow and transport in complex systems. They recognized the importance of not being seduced by broad generalizations about all networks based upon appealing patterns in a single one.

"It is important to remember that the unique subtleties of individual networks and the dynamics along that network ... affects the means of control along that network," the group agreed.

The other half of Group 9 explored damage control in complex systems. In the wake of the recent economic collapse, the group sought to use Google-search red flags, as has been done with flu outbreaks, to warn of coming economic collapse. Group members postulated that two forces govern information and financial flow—gain and fear—and that such a warning system could alert those in control of interest rates and other relevant financial data to rapidly react when the system switched from pursuit of wealth mode to avoidance of loss mode.

Taken as a whole, the challenges are both mechanical and epistemological, both chemical and philosophical, and the questions are the ones that will define us as a species within our ecosystem. From fighting disease to reversing environmental damage, the quest to effectively model our bodies, our social groups and our effects on the planet is a profoundly important one. As explorers, we must seek to replace the indistinct regions on our maps with meaningful topographies, and in so doing, better know ourselves.

Task Group Summary 1
How would you design the acquisition and organization of the data required to completely model human biology?

CHALLENGE SUMMARY

In many fields that relate to complexity, datasets are still fragmentary and questionable in terms of their overall quality. This is particularly true in the field of biology. Small-scale empirical data have been described for decades in hundreds of thousands of papers published in thousands of journals. This information, although generally perceived as highly accurate, is extremely hard to extract in reliable ways. On the other hand, high-throughput systematic biological datasets tend to be widely accessible, but are currently perceived as lesser quality information. This represents a considerable challenge if one considers the fact that, relative to its widely accepted complexity, the molecular aspects of human biology have been described only superficially.

KEY QUESTIONS

With the general assumption that we are given funding in the range of what was allocated to sequence the human genome between the late 1980s and the early 2000s (~$3,000,000,000), the following questions will be addressed:

- How would you *design* the acquisition of new data pertaining to human biology?
  - How would you *validate* the inherent quality of such data?
  - How would you organize this information into practical, usable

datasets made available in databases ready to be used by the research community?
- How would you design the development of analytical tools to attempt to entirely model the molecular and physiological complexity of the human body?
- How would you relate this information with genetic and environmental factors that influence disease and good health?

Required Reading

Aloy P, Russell RB. Potential artefacts in protein-interaction networks. *FEBS Lett* 2002;530:2556.

Brazma, et al. Minimum information about a microarray experiment (MIAME)—toward standards for microarray data. *Nature Genetics* 2001;29:365-71.

Fields C, et al. How many genes in the human genome? *Nature Genetics* 1994;7:345-346. Editorial. *Nature Genetics* 2000;25:127-128 **and references therein.**

Maslov S, Sneppen K. Specificity and stability in topology of protein networks. *Science* 2002;296:910-913.

Maslov S, Sneppen K. Protein interaction networks beyond artifacts. *FEBS Lett* 2002 Oct 23;530(1-3):253-254.

von Mering C, et al. Comparative assessment of large-scale data sets of protein-protein interactions. *Nature* 2002;417:399-403.

Yu, et al. High quality binary protein interaction map of the yeast interactome network. *Science* in press.

Suggested Reading

Noble D. The music of life. Oxford: Oxford University Press 2006.

TASK GROUP MEMBERS

- Ananth Annapragada, University of Texas Houston
- James Glazier, Indiana University
- Amy Herr, University of California, Berkeley
- Barbara Jasny, Science/AAAS
- Paul Laibinis, Vanderbilt University
- Suzanne Scarlata, Stony Brook University
- Gustavo Stolovitzky, IBM Research
- Eric Schwartz, Boston University

TASK GROUP SUMMARY

By Eric Schwartz, Graduate Science Writing Student, Boston University

The question of how to put together and organize the data needed to simulate human biology is large and complex. At the 2008 National Academies Keck *Futures Initiative* Conference on Complex Systems, a Task Group (1) of scientists from multiple disciplines met to contemplate the problem.

The goal is a complete, easily queried simulation that would be comprehensive and could synthesize different data to give useful answers to questions about human physiology in health and disease. This is obviously a monumental undertaking, especially when we realize the limitations of current state-of-the-art computers and technology, and our mental ability to conceptualize such problems. Nevertheless, the group developed an initial plan upon which many future directions can be based.

First, there is the challenge of obtaining information that scientists know would be essential. For instance, it is estimated that humans have approximately 25,000 genes, but the interactions of only about 10% of their products are known. Proteins made by genes, in multiple forms, interact with each other in different ways. A comprehensive simulation might require knowledge of protein production and behavior in space and time. (Not all proteins are active at all times or at all places.) Metabolic, signaling, and gene regulatory pathways, known and yet to be understood, would be part of the simulation, as would patterns of neuronal growth and decay, and whole organ anatomy and function, and nervous, endocrine, circulatory, respiratory physiology systems, etc. Altogether, simulating human biology is an immense problem not only of biological research but also bioinformatics, biomedical computation, epistemology, and computer power. Unless all of this information is combined in an understandable format, a lot of important and relevant medical data for a human simulation would be ignored.

The Initial Plan

The group considered many options for collating and organizing data. It was decided that one of the most important steps is to find out what empirical data compilations already exist and organize them according to some basic principle to avoid covering ground already covered on a scale ranging from the molecular, protein, cellular, organ, and full-organism scales. There may be more than 250 types of cells in the human body, each with their own unique functioning and therefore the group decided the interaction between different levels of biology is just as important as what occurs on those levels alone. Ultimately, correlation of the different types of data rests on good indexing, or a metastructure to define all of the phenomena. The group concluded that the best way to deal with the question of human biology as a whole is to break it up into different parts. Five basic databases were outlined by the group with the expectation that the databases could then interact with each other. The databases enumerated were:

1. Simulations of subsystems and connections. In this case meaning cellular, protein, and other systems and how they relate to each other.
2. Limitations: that is, the limitations posed by the lack of standardized datasets on human biology and the ability to relate these data to information about biological simulations.
3. Experimental data plus metadata for different cases and perturbations.
4. Templates of appropriate subsystem choices and connections for different categories of problems.
5. Sample complete simulations and outputs.

The group then broke down the databases into smaller subsets of knowledge. The database of parameters was for example further subdivided into spatial and temporal distribution, mechanical properties, cell behavior, and biomarkers. The group decided the most important issue facing them was the many gaps in their knowledge. The different databases currently in existence aren't standardized and there is no consensus ontology or unified computational tools to deal with the data already compiled. Ontologies are logical structures which provide a formal description of concepts. An ontology is simply a hierarchy of terms with understood meanings and sets of subterms and modifiers which can be applied to each term.

In order to know what to do with the overwhelming complexity of the whole of human biology, the group decided that a pilot program to test their basic ideas is essential. If successful, the pilot simulation could then be the basis for simulations of other parts of human biology. The pilot would need to be something simple but at the same time useful for discussion. Although many options, from cancer to neurodegenerative diseases, were considered, the group settled on the effects of an injection of norepinephrine into the body. This chemical is used on people in anaphylactic shock caused by allergies or toxins and has several well-understood effects. By comparing what is known about norepinephrine in people to the theoretical simulation, the simulation could be tested for predictive capacity.

The Five Year Plan

The group resolved to create a list of goals that could be achieved within five years, should sufficient resources be applied to the work of a complete simulation of human biology—"Google Human." Firstly, they wanted to create an inventory of all the data currently available and a preliminary inventory of all the missing data. Once the data have been created and compiled, a quality control check of all the data will be necessary to make sure that the data are correct and put into a format that is consistent for computer analysis. Along with creating standards, designing new computational tools will be an important early step in the program.

For a detailed outline of the group's thoughts, see Figure 1.

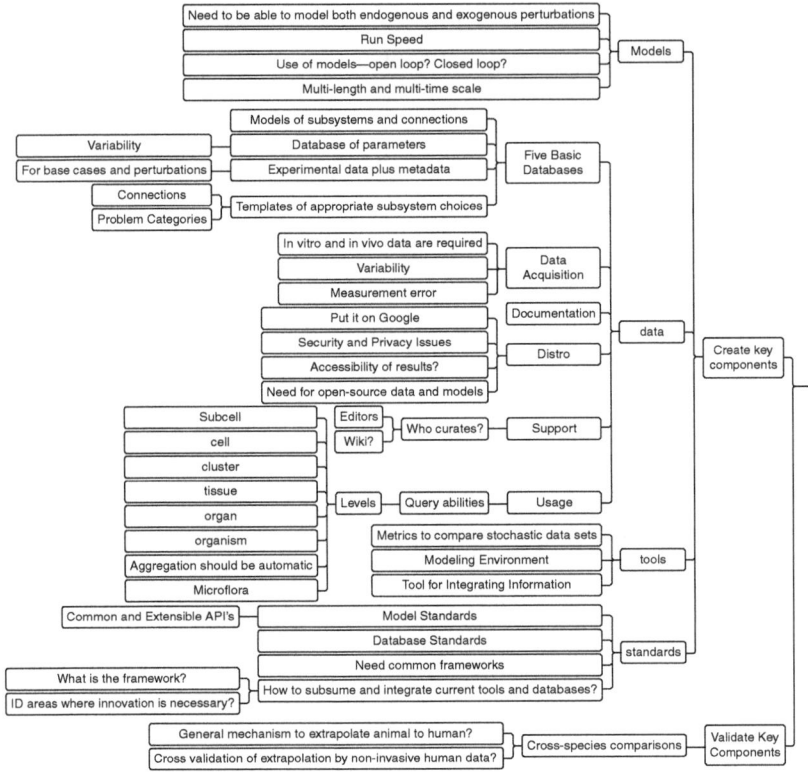

FIGURE 1 The Initial Plan

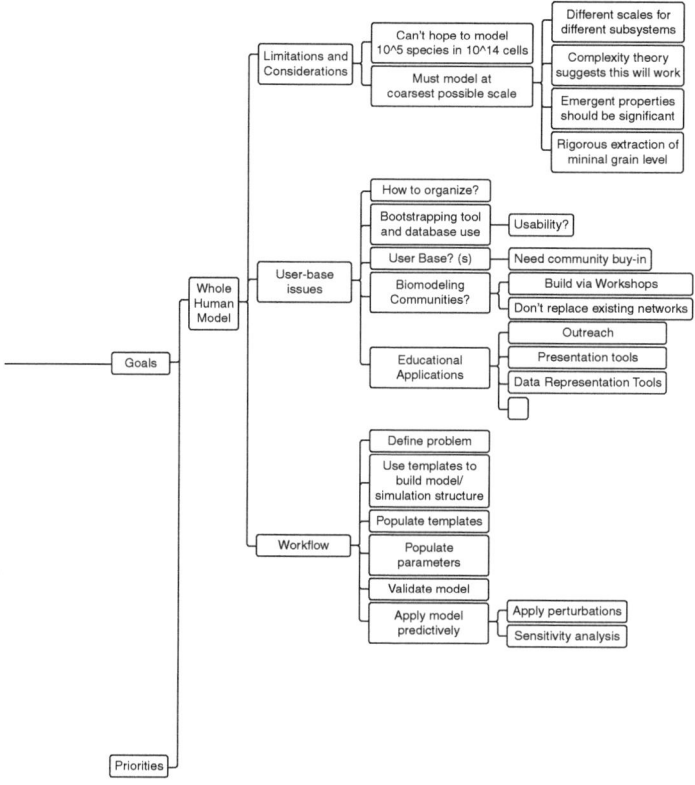

Task Group Summary 2
What does it take to achieve a sustainable future? The problem of the commons: achieving a sustainable quality of life.

CHALLENGE SUMMARY

Eight hundred million people are chronically hungry today and perhaps another 100 million will be chronically hungry within a year. Sadly, this statistic reflects no inability of the Earth and humans to produce enough food for all. We grow enough cereal to adequately feed a population of 10 billion people (current world population 6.6-6.7 billion). Issues in hunger include:

- Consumers' inability to pay for food
- Growers' inability to purchase seed, fertilizer, equipment, and other necessities for growing food, to get it to market, and to sell it at a profit
- Cultural taboos such as fear of genetically modified organisms
- Cultural limitations on what foods people are willing to eat
- Politically motivated agricultural subsidies in rich countries that undercut the ability of food producers in poor countries to compete in world markets
- Food price controls imposed by fearful governments in developing countries that limit farmers' incomes

Demographic projections suggest that nearly all of the next 2.5 to 3 billion people to be added to the planet by 2050 will live in cities in poor countries. That projection, if correct, requires building the equivalent of an additional city of 1 million people every five days for the next 40 to 50 years. Are the problems of assuring decent health, adequate sanitation, housing, food supplies, amenities, and public order for such cities soluble with present scientific and technical knowledge? If not, what do we need to do to prepare for this growth?

Given these challenges ideas for a sustainable human population must be developed. Can we learn from simpler model systems, such as microbial systems in which enormous numbers of like and unlike individuals can be grown and subjected to different environmental and nutritional stresses? Perhaps such simple systems can be modeled and the outcomes experimentally verified. Particularly hospitable and particularly inhospitable situations can be examined.

In addition, we could examine evolutionary theory, particularly theories of social evolution and mechanisms of conflict and conflict resolution be used on a global human scale.

Key Questions

- How can we deal with problems of the Commons, in which the collective consequences of individual behaviors affect public goods? Can we model these interactions between units with dynamics at such different organizational and spatial scales? Can an understanding of collective phenomena and cooperation in non-human societies inform better stewardship of resources and our common environment (Levin 1999)?

- Can we meaningfully describe the complexity of any population by a few factors (for example: population, resources, environment)?

- There is evidence that desired family size is affected by cultural norms and has a contagious or imitative effect. It has also been clearly demonstrated that external phenomenon, such as the empowerment of women through education and employment, leads to smaller desired family sizes, as the "extra hands" motivation diminishes. How could these factors be best combined to effect population control policies, while maintaining individual freedom and equity?

- Can we use tractable experimental systems to help us learn what to expect as the human population increases in a warming environment?

- Can theories derived from the field of social evolution aid us in understanding and predicting human population changes?

Required Reading

Arrow K, Bolin B, Costanza R, Dasgupta P, Folke C, Holling CS, Jansson BO, Levin S, Maler KG, Perrings C, Pimentel D. Economic growth, carrying capacity and the environment. *Ecological Applications* 2006;6(1):13-15.

Commission on Growth and Development Study 2008. [Accessed June 10, 2008: http://www.growthcommission.org/index.php?option=com_content&task=view&id=96&Itemid=169.]

Suggested Reading

Cohen J. How many people can the earth support?. New York: W.W. Norton & Company, Inc. 1995.

Levin S. Fragile dominion: Complexity and the commons. New York: Perseus. 1999.

TASK GROUP MEMBERS

- James Crutchfield, University of California, Davis
- Ana Diez Roux, University of Michigan
- Doyne Farmer, Santa Fe Institute
- James Gardner, Gardner & Gardner, Attorneys, PC
- Murray Gell-Mann, Santa Fe Institute
- Jessica Hellmann, University of Notre Dame
- Paul Humphreys, University of Virginia
- George Kaplan, University of Michigan

TASK GROUP SUMMARY

By Monica Heger, Graduate Science Writing Student, New York University

If the entire world consumed as much as the average United States citizen, we would need 4.5 Earths to sustain us. And, if we continue to grow and increase consumption at the same rate, by the 2030s we will need two planets to support our way of life, according to *The Living Planet Report 2008*, an annual assessment produced by the World Wildlife Fund, Zoological Society of London, and the Global Footprint Network. These figures suggest that humanity is on course for environmental collapse, and a new, more sustainable future must be implemented if we are to forestall that collapse.

Experts in biology, physics, philosophy, economics, anthropology, public health, and law, all came together at the 2008 National Academies Keck *Futures Initiative* Conference on Complex Systems to examine the problem of sustainability, describe what a sustainable future would look like, and predict whether or not it is in fact possible.

The group decided that before a sustainable solution can be drafted, we need a modeling framework that explores possible future outcomes. A new model that the group named SOS World, for scientific open source, would be a freely available model of the world that would take into account climate, economics, demographics, health, and any other input deemed necessary to simulate what the future could have in store for us. The group envisioned large groups of scientists contributing to the development of SOS World, a scientific tool that would inspire solutions and serve as a platform for developing sustainability as a research effort.

What Is Sustainability?

Many traditional economic models assumed that humanity depends on steady economic growth worldwide. Yet, as demonstrated by the current financial crisis, endless economic growth is probably not possible, nor is it necessarily desirable. Just because a country's GDP is growing, does not mean that the country is living sustainably, nor does it equate with increased quality of life for all of its citizens. The group agreed that economic growth should not be the sole measure of sustainability and that for the purposes of its analysis, sustainability would be loosely defined as quality of life not purchased at the expense of the future. Sustainability may or may not be possible, but the group said it is nevertheless necessary to work towards a future that comes as close as possible to sustainability. The primary goal of SOS World would be to identify sustainability when it arises in the model, rather than prescribing what it would look like in advance.

To move towards sustainability, seven key changes need to occur:

1. Demographic—away from unchecked population growth
2. Technological—develop and implement technology with less environmental impact
3. Social—towards social equality
4. Economic—to more economic equality
5. Institutional—to more effective means of coping with conflict
6. Informational—to greater access to information and education
7. Ideological—from ideologies that divide, to ideologies that unite

A sustainable world would be one that resulted in changes in all seven of these areas because in order to engage citizens across the world in tackling some of the greatest environmental problems, everyone would need to have the resources and opportunity to participate. Developing countries will not have an incentive to participate if doing so does not also improve their standard of living. Educating people in developing counries is also important because they could very well have something important to contribute.

A Complex Systems Model Based on Neural Networks

In thinking about how to create a model that encompasses many different variables—health, economics, climate, ecology, education, and others—it is necessary to use a model that represents a complex system; in this case a neural network. Neural networks are modeling techniques that can analyze nonlinear systems. They were originally developed to explain the neural networks in the brain, but can be applied to almost any system that cannot be explained linearly. They can be used in systems where there are multiple inputs and where there is a relationship between the inputs and the predicted outcomes, even when those relationships are complex. So for something like an ecosystem, that is affected by many different factors, a neural network can analyze those very complex relationships and simulate the interaction among dozens if not hundreds of agents and processes.

The other main advantage of a complex systems model is that it can identify tipping points—points where small changes have huge effects on the system. These tipping points will be useful in shaping policy because they will identify crucial areas where a small change for the worse could have devastating consequences. Alternatively, it will also pinpoint small changes that will have a large positive impact, which will help create sustainability in a cost effective way. In this way, sustainability will be defined as outcomes that emerge from simulation that achieve all or many of the goals above.

An SOS World

The resulting model would be one that could be replicated and modified as necessary. The SOS World model would take into account any and all factors that are deemed important to achieving sustainability, including climate, economics, population growth and health. Sustainability can be viewed in more than one way, or as encompassing different systems, which in turn could lead to different predictive outcomes. Some scenarios or possible outcomes that SOS World might generate, depending on the parameters in the model, include:

- Growth World – Never-ending economic growth and consumption everywhere all the time. This was the world as described by some of the experts who use economics as the only measure of sustainability. The group decided this was not a realistic or desirable world.
- Death World – Everything declines. The group seemed to think this was a real possibility, particularly if the status quo is maintained.
- Wave World – Growth and prosperity move in waves across the globe.
- Chaos World – Random fluctuations of growth and prosperity. This would be similar to booms and busts, like the housing and tech booms and busts.

The purpose of creating the simulations is to identify the important variables that need to be changed to create a world that resembles sustainability. Even though the model will take into account many different variables, by using a complex systems analysis, such as a the neural networks analysis, it will be possible to identify crucial tipping points and make meaningful policy based on those tipping points, rather than piecemeal policy that often has unintended consequences. In short, the model is assembled with important parts of the global system, and the outcomes represent possible future trajectories. A key question is: Can a sustainable outcome emerge, and under what conditions (parameters) does it emerge?

Next Steps

The group intends to pursue its work at NAKFI by applying for a grant to design the model platform, recruit experts, and see what other work is being done in the area in case it is possible to combine efforts. Realizing that the project is a huge undertaking, the scientists also plan to seek out long-term funding, particularly to craft the design platform for SOS World.

The enormity of the task of first modeling and then creating a sustainable environment was not lost on the scientists, who nonetheless were driven by the urgency of the challenge. The group's adopted mantra—if not now, when?—sums up nicely the sense of responsibility and necessity the group felt towards solving the problem of sustainability. How do we live in a sustainable way, how do we ensure our children and grandchildren have a future, how do people living in poverty grow out of it? And how do we devise a model that can meaningfully address all these questions?

Task Group Summary 3
How can we enhance the robustness via interconnectivity?

CHALLENGE SUMMARY

In contrast with most human designs, which are prone to failures once their components fail, natural and some human-made but self-organized systems display a high degree of robustness to component failures. Indeed, living systems can carry on their activity despite the many molecular errors at the cellular level and the Internet does not collapse despite the fact that at any moment hundreds of routers are not functional. Many living systems, like bacteria, have been shown to be able to withstand the removal of several key enzymes. It is increasingly believed that the robustness of these systems is rooted in their networked nature. Early attempts to address a network's response to attack and failures indicated that real networks are highly robust to random failures but fragile against attacks. Subsequent work has shown that the interplay between the resources and the demand can lead to cascading failures, uncovering a high degree of fragility of some systems. A good example is offered by the US electrical power grid, whose cascading East Coast breakdown was initiated by local failures. In general, a series of recent studies suggest that networked systems are not only robust but also suffer from *vulnerability due to interconnectivity,* as local failures can spread and turn global.

Despite the recent fundamental advances, a deep understanding of the origins and mechanism or robustness across many complex systems is lacking. Little is known, for example, of the role of the dynamics (communication protocols, flow processes) on the network, and how the dynamics and the topology influence each other to promote or undermine robustness.

Thus the role of the present working group is to explore what factors contribute to a system's robustness. To achieve this goal, the group is asked to choose a system that is of major importance for the research community and explore the origins of robustness in this system. The system of choice could range from man-made systems, like the Internet or other communication networks, to natural systems, like the cell or an organism.

Key Questions

- What are the proper metrics of robustness?
- How does one quantify the relative contributions of network structure and dynamical effects to robustness?
  - Are there universal design principles to robust systems?
  - Is robustness more than redundancy?
  - Designing networks that are robust to both failures and attacks.
  - Cascading failures—can they ever be remedied?
  - What measures are appropriate to enhance robustness on a given system?

Required Reading

Albert R, Jeong H, Barabási A-L. Error and attack tolerance of complex networks. *Nature* 2000;406:378–482.

Motter AE. Cascade control and defense in complex networks. *Phys Rev Lett* 2004;93(9):098701. [http://lanl.arxiv.org/PS_cache/condmat/pdf/0401/0401074v2.pdf.}

Barabási A-L, Bonabeau E. Scale-free networks. *Scientific American* 2003:May:50-59.

Suggested Reading

Levin SA, Lubchenco J. Resilience, robustness, and marine ecosystem-based management *Bioscience* 2008;58(1):27-32.

Paul G, Sreenivasan S, Havlin S, Stanley HE. Optimization of network robustness to random breakdowns. *Physica A* 2006;370:854-862.

Barkai N, Leibler S. Robustness in simple biochemical networks. *Nature* 1997;387(6636):913-917.

Due to the popularity of this topic, two groups explored this subject. Please be sure to review the second write-up, which immediately follows this one.

TASK GROUP MEMBERS – GROUP A

- Ramanand Dixit, Washington University in St. Louis
- Rebecca Goolsby, Office of Naval Research
- Natali Gulbahce, Northeastern University
- John Hartman IV, University of Alabama at Birmingham
- Pradeep Kumar, Rockefeller University
- Arthur Lander, University of California, Irvine
- Christopher Myers, Cornell University
- Aristides Requicha, University of Southern California
- Qian Wang, University of South Carolina
- Casey Rentz, University of Southern California

At the 2008 National Academies Keck *Futures Initiative* Conference on Complex Systems, one of two Task Groups (3A) charged with thinking about how to enhance robustness via interconnectivity, considered several areas on which to focus. Group members from universities and government research centers saw complex systems from a variety of different perspectives: nanoscale bond interactions; microtubules and systems of self-organization in cell growth; sensor networks in the coordination of robots; human disaster relief social networks; water and turbulence; virus life cycles in the human body; and gene interaction networks derived by quantitative phenotyping.

What Is Robustness?

The group initially grappled with what we mean by robustness. The general consensus was that we needed to define what the system is before we talk about its robustness. For example, is the system a cell, an organism or species? Because there are many facets to complex systems across different scale-levels, different perturbations and performance measures need to be considered. One must be careful of what scale is selected when initially defining the complex system (and related optimization goals). In certain systems, one might find performance fluctuations on a small scale that would not appear at a larger scale. Or, a loss of robustness at a certain scale might be accompanied by a gain in robustness at another scale. These robustness trade-offs, where performance or robustness of a system is sacrificed at one level to be enhanced at another level, exist and must be taken into account when robustness of a complex system is measured. A complex system is also typically fluid. Connections within an engineered or biological complex system are always breaking, reforming, and changing. Perturbances often seem inseparable from the networks and complex systems themselves.

Everyone in Task Group (3A) seemed to agree that robustness is a continuum, not a case of have or have not.

Exploring the nature of robustness and fragility in complex systems, Task Group (3A) attempted to abstract from real-life examples of complex systems how connectivity within the system might influence its robustness.

"Networks" provide a useful way to depict a complex system through component nodes and functional connections. For convenience of discussion, the group identified four systems that were deemed relatively easy to deconstruct into their component parts: power grid structures, health care, the Internet, and yeast genetic interactions. For example, power grids have as nodes houses/businesses, substations, and central power stations; highly connected nodes (central stations) are considered as hubs. Similarly, nodes of a health care system could consist of patients, doctors and other health providers, connected by their respective encounters. The Internet can be broken down into personal computers as nodes and servers as hubs of information distribution. A genetic interaction network has genes as nodes, and connectivity between the nodes represent "interactions," defined as dependencies that genes share with respect to expression of a phenotype, like cell growth.

An abstract examination of network dynamics and degree of interconnectivity within the structure of our selected complex systems allowed us to make recommendations for increasing robustness. A network's behavior can be thought of as based on its performance in a particular context due to the effect of a particular perturbation. We delineated hypothetical performance objectives and relevant perturbations for selected complex systems, with an eye toward abstracting general robustness strategies from one system that could be applied in an analogous way to increase robustness in other systems.

In the case of power grids, the objective in enhancing robustness was to maximize the number of people with electrical power and to minimize the risk of cascading power failure. In the case of health care, the objective was to prevent epidemics caused by a novel pathogen. In the case of the Internet, the objective was to maximize available online time for each individual while preventing service failures. In the case of yeast genetic interactions, the objective was to characterize robustness by "reverse-bioengineering;"

cell proliferation is a robust property of cells based on the observation that individual deletion of most yeast genes has little effect on growth. However, high throughput phenotypic analysis of cell proliferation of all 5000 gene deletion mutants in many different types of media provides a systematic, quantitative means to ascertain how genes contribute to the cellular robustness and ability to adapt to changing environments.

Interconnectivity and Dynamics

Imagining the effect of a particular perturbance, as it would spread throughout each model complex system, it is clear that interconnectivity can lead to both robustness and vulnerability. For example, "essential genes" (deletion results in lethality) have a greater number of physical (protein-protein) interactions than non-essential genes, and thus can be considered as cellular "hubs" of function. Likewise, the more interconnected a power substation, the more people it supplies, but it is also an easy target for a blackout. A perturbance, such as a novel disease, would spread through and weaken a complex system, such as the health care system, through nodes that are highly connected. But spread of a perturbance is also greatly affected by system dynamics. For example, a sick individual, one node of the health care system, may have a particularly virulent strain of a specific disease (or be somehow better at transmitting the disease). Though they are not well connected to the rest of the system, the disease would pass through the system more quickly due to the strength of this nodal connection. Additional factors that can affect network dynamics are directionality and strength of links; both of which can be used to determine where a perturbance will flow and where vulnerabilities in the system will arise.

Knowledge of directionality and strength of interconnectivity can potentially be exploited to increase robustness. For example, such knowledge would allow us to "park" fragilities where they are least vulnerable and most easily managed. A simple example of this is power stations, which as sources of electrical power in the power grid system should be the most protected hubs in the system in order to lessen the consequences of a malfunction or attack that would otherwise result in a catastrophic breakdown of the system. In the case of genetic interactions, understanding fragilities that result from cancer-causing mutations would reveal targets for selectively killing cancer cells. Control in a complex system does not necessarily have to coincide with hubs of that system. In most dynamic complex systems, blending of centralized and distributed control would enhance robustness.

In addition to active control, adaptability of system topography increases robustness in a complex system. The group discussed random and scale-free networks as two kinds of network topologies that respond differently to perturbations/attacks and would thus affect robustness. A random network is one where the nodes are equally interconnected and/or evenly distributed throughout the network while a scale-free network has a few hubs that are highly interconnected, with the majority of nodes having fewer connections.

One way to enhance robustness in a complex system is to create a topology that is an adaptive mix of a random and scale-free network. With some knowledge of the kind of an attack or perturbation to a complex system, the system would be able to switch states depending on the nature of the perturbation. For example, if substations of a power grid were attacked, the system could switch topologies and begin to distribute power evenly through all its nodes, switching from a discrete to diversified network. This adaptability increases robustness of the system, but would require a tradeoff in expense to implement and maintain necessary resources.

The group also discussed vertical connections and redundancies as attributes of a complex system that could contribute to robustness of the system. Some biological systems are among the most robust complex systems in existence. Experimental data from yeast genetic interaction experiments, indicate, for example, that simple redundancy of function accounts for a small amount of the observed robustness. It seems "alternative pathways" and dynamic rerouting of system fluxes in response to perturbation are often the adaptive mechanisms that contribute to robustness in biological systems. "Vertical connections" are also very important in robustness considerations. These are connections in a network of a complex system that span different hierarchical levels, scales, or employ different definitions within a sub-system.

Additional mechanisms are required of a system to maintain and enhance robustness. These include feedback mechanisms in networks, self-organization and self-repair mechanisms. Feedback is also important for establishing buffering mechanisms that increase robustness by stabilizing a dynamic system against perturbations. A simple example of this would be Internet servers. Often, if one server is down, information gets passed to another server in a cluster so that Internet clients can still access information though their home computers. This buffering mechanism ensures that clients have maximal access to the Internet at all times. Gene-gene and gene-environment interactions reveal buffering relationships through analysis of combinations of perturbations that are synergistic or antagonistic with respect to cellular robustness. Self-organization and self-repair are also present in biological systems and are beneficial for enhancing robustness in any complex system.

Moving Forward

As the group moved from thought experiments, and attempted to extrapolate findings into the real world, it became possible to generally but solidly define the task of seeking robustness in a system as:

Keeping the magnitude of the change in a set of performances (with respect to a set of perturbations) within some limits, and doing so subject to given restraints.

Definitions of complex systems in real life can be difficult. There is a hierarchical structure of biological systems such that different measurement scales apply to different levels of the system; different types of perturbations are relevant, and different performance measures must be considered. One must be careful of what scale is selected when initially defining the complex system (and related optimization goals). In certain systems, one might find performance fluctuations on a small scale that would not appear at a larger scale. Or, a loss of robustness at a certain scale might be accompanied by a gain in robustness at another scale. These robustness trade-offs, where performance or robustness of a system is sacrificed at one level to be enhanced at another level, exist and must be taken into account when robustness of a complex system is measured.

Defining a complex system is also difficult due to inherent dynamics. Connections within an engineered or biological complex system are always breaking, reforming, and changing. Perturbances (e.g., signal transduction) often seem intrinsic to the networks and complex systems themselves.

It is challenging to study a robust system that by definition consists of dynamic interactions rendering it resistant to observable change. Despite this challenge, we can still make recommendations for enhancing its robustness. Robustness is incremental and non-linear, so we need to establish quantitative models and tools to measure sources of buffering capacity and better model phenomena such as stability thresholds. Adaptability of network topology within system structure is also key to enhancing robustness, as are built-in feedback mechanisms and active control.

When enhancing robustness or minimizing vulnerability, it is highly advantageous to know as much as possible about the interconnectivity and dynamics of a system. For example, if biological attack were to be mounted within our health care system, it would be advantageous for the attackers to know which individuals are the most connected, and would spread the disease most quickly. This knowledge could be used to attack fragilities as well as hide them, to decrease or enhance robustness of this system. System dynamics, along with interconnectedness, allow prediction of how the system will function and respond. Analysis of diverse types of complex systems, such as biological and man-made systems, is an important aspect of our aspiration to derive principles for how robustness is achieved through network structure and connectivity.

TASK GROUP MEMBERS – GROUP B

- Kirstie Bellman, The Aerospace Corporation
- Jennifer Couch, National Institutes of Health
- Raissa D'Souza, University of California, Davis
- Tony H. Grubesic, Indiana University
- Stephen J. Kron, The University of Chicago
- Luis Rocha, Indiana University
- Caterina Scoglio, Kansas State University
- Alejandra C. Ventura, University of Michigan
- Stuart Fox, New York University

TASK GROUP SUMMARY – GROUP B

By Stuart Fox, Graduate Science Writing Student, New York University

Fragility is an inherent component of all systems. But unlike a simple system, in which fragility is equally distributed, complex systems present an uneven landscape of strength and weakness. As a result, robustness, the ability of a system to limit, within some specified range, the magnitude of change in performance with respect to perturbations, can only be understood, enhanced, or engineered with the proper intellectual framework. Researchers at the 2008 National Academies Keck *Futures Initiative* Conference on Complex Systems were unable to develop a complete version of that intellectual system; nonetheless, the Task Group (3B) successfully identified the key questions that the intellectual framework would need to answer, and began the process of constructing that framework.

The first step in understanding robustness is acknowledging that robustness is context dependent. This quickly became apparent as the group attempted to answer its original question, which focused on interconnectivity and robustness. In some cases, like a gene regulatory network, more connections mean more robustness. But it is not hard to imagine a reverse scenario, like increased plane travel helping spread a deadly pandemic, where interconnectivity decreases the robustness of a system.

It turns out robustness does not result from finite set of qualities, the application of which could steel any system against failure, but instead depends on the system in question, the goals of the system and the perturbations that system faces. Different strategies will work for some systems and not others. Robustness may mean preservation or adaptation depending on the system, and there is always a cost.

The cost may come in the form of money, metabolic energy, loss of robustness against a different set of perturbations, and many more forms. In fact, the costs of robustness are as numerous as the strategies for creating and enhancing it. The costs of robustness can never be eliminated, only shunted from one area of the system to another.

The very dependence of robustness on context naturally suggests a vague outline of the intellectual framework needed for the understanding of robustness in specific examples, and for the engineering of robustness in man-made complex systems. That framework, at least insofar as the group was able to divine, requires asking four key questions: What is the goal of the system? What are the perturbations? What strategies preserve the goal of the system in the face of those specific perturbations? And what is the cost of each strategy?

To test the elegance of those questions, the group used that preliminary framework to analyze four model systems (see Table 1). Looking at each system, whether natural or man-made, the group found that the robustness strategies lend themselves to grouping far more naturally than the goals of the systems, the perturbations, or the costs. This early result hints that while perturbations and costs vary as widely as complex systems themselves, there may be a finite set of robustness strategies applicable in different situations (see Table 2).

The first, and simplest, systems the group looked at were the proteins RNAse A and green fluorescent protein (GFP). RNAse A is an enzyme that cuts up RNA molecules in liver cells and serves as the standard biochemi-

TABLE 1 Example Systems

System	Perturbation	Strategy	Cost	What Is Preserved
RNAse A, GFP	Temp, pH, cutting into piece	Self-healing, redundancy	Failure to adapt, design time	Enzymatic function, glowing
Central and peripheral nervous system	Temp, chemicals imbalance, tumors, stroke, blow to the head, etc.	Hardening, diversity, avoidance… everything in Table 2	Metabolic energy (vastly disproportionate consumption), design time	Life, cognition, ability to adapt
Federal highway network	Congestion, blockages, link failure	Diversity of paths, spatial separation, provisioning	Construction, maintenance	Transport, origin-destination paths, flow
Gene regulatory network	Mutations, environmental conditions, development stages	Canalization, connectivity, redundancy, multitasking	Constructing additional molecules, design time	Phenotype, multiple viable modes (adaptability)

TABLE 2 General Robustness Strategies

• Redundancy (repetition, substitution, overlap)	• Feedback
• Diversity	• Smart nodes/smart edges
• Modularity	• Self-healing
• Spatial separation	• Regeneration
• Fortification	• Balancing control between centralized and decentralized components
• Buffering	
• Cutting losses	
• Canalization	
• Avoidance	• Trusted/collective intelligence for centralized aspects
• Connectivity	
• Multiple viable modes	
	• Computational reflection (system's awareness and planning of resource allocation)

cal model for protein studies. It gained that model status when Armour & Co., the company that makes Armour hotdogs, purified 10 kilograms of the material and distributed the enzyme for free to research institutions. GFP is a luminescent protein found in jellyfish that is widely used as a marker for biological processes in experiments.

Both proteins are far more robust than other similar proteins in the face of denaturing as a result of heat or pH, being cut up, and to the replacement of component amino acids with other amino acids. When the proteins are denatured and the perturbations of heat or pH are removed, or the cut up parts of the proteins are put together and heated, the proteins spontaneously reform. If they are improperly assembled, the proteins still retain some functionality.

That robustness comes at the cost of adaptation. The robustness focuses on maintaining a consistent form, preventing the proteins from undergoing any changes that might enhance the robustness of the protein in the face of other perturbation.

Another biological system the group examined was the nervous system, both central and peripheral. The central nervous system consists of the brain and spinal cord, and controls cognition. The peripheral nervous system connects the brain to the rest of the body, and controls all other functions. The nervous system faces perturbations such as chemical imbalance, change in pH, loss of oxygen, prion disease, stroke, and blunt force trauma, among others.

To deal with those problems, the nervous system uses nearly every robustness strategy known to science. For example, a hard skull protects against trauma, an immune system protects against disease, redundancy protects against stroke and single channel failure, flight or fight response avoids or defeats threats to the system.

The cost of that wide spectrum of robustness comes in the form of metabolic energy and design time. The brain takes up most of the energy in the body, and a lot of food is needed to fuel a brain that can think its way around perturbation. As far as design time goes, it took evolution hundreds of millions of years to progress from the simplest nerves to the human brain.

Of the man-made systems that can be analyzed for robustness, the group spent the most time examining the federal highways. The government created the highway system in the 1950s to ensure cross continental military supply lines would not be interrupted by Soviet nuclear attack. As the threat of nuclear war receded, the highway system shifted to primarily providing travel arteries for citizens and companies. In both cases, traffic congestion, blockages and link failures were identified as problems.

Making sure people, products and tanks get where they are going in the face of those problems requires multiple paths to the same destination, increased linkage between the roads of the federal highway system and state and local highways, and geographic distance between hubs to protect against catastrophic, regional problems like hurricanes and nuclear explosions.

Robustness of the highway system has both straightforward and more nuanced costs. The obvious cost is money. Construction and maintenance of the highway system need to be funded. The more subtle cost arises through a concept called Braess' paradox. It turns out that the robustness strategy of adding more roads to decrease congestion actually leads to more traffic in some cases.

While these examples set the group on the right track, many questions remain unanswered. How does one generalize strategies from examples? How best to implement a general robustness strategy with the proper specific detailed information to make it work on the target system? What benefits and costs do different general robustness strategies provide for specific complex systems? Moving forward, the group decided to continue working on this problem, and over the course of time, to finish developing the framework for understanding, leveraging and enhancing the robustness of complex systems.

Task Group Summary 4
Can engineering systems and control approaches generate new strategies for altering imbalanced macrophage profiles in human disease?

CHALLENGE SUMMARY

Biological organisms are complex systems whose design has been shaped by billions of years of evolution. We are now at the threshold of identifying the entire "parts list" (genes, proteins, metabolites, etc.) for many organisms including humans. The enormous task confronting us currently is to understand how these parts work together in complex hierarchies of systems within systems. The only analogous systems whose complexity we can claim to understand are complex engineered systems.

The reductionist approaches of cell and molecular biology have been spectacularly effective in revealing the elements of biological design, but these approaches are incapable of providing the quantitative and integrative techniques that are required to reveal the design principles of the intact system, the repertoire of its dynamic behaviors, and its pathologies.

There is an emerging focus of activity on discovering biological design principles that draws upon our experience with complex engineered systems. This experience is providing a useful guide for this discovery process. With this understanding one can begin to envision rational strategies for reengineering biological systems for therapeutic purposes. For some time complex biological systems will receive the most benefit of this focused activity at the interface between biology and engineering. However, the design principles of biological systems are likely to suggest new ways of providing robust control of complex distributed systems that will also benefit the design of complex engineered systems.

The specific challenge for this task group is the increasing recognition that immune responses involving macrophage cells play important roles both in protection against disease and in the promotion of disease (DeNardo & Coussens, summary in Fig.3, 2007). Characterizing the phenotype, molecular signaling, and therapeutic opportunities associated with these warrior cells has great relevance in the diagnosis and treatment of conditions ranging from vascular disease to cancer. Two distinct phenotypes have emerged in the systems analysis of cancer—an "M1" phenotype associated with acute inflammation and tumor rejection and an "M2" phenotype associated with chronic inflammation and tumor progression. The distinct phenotypes can be identified by their specific signals: secretion of anti-tumor cytokines in M1 and secretion of tumor-promoting growth factors in M2.

Key Questions

The challenge to the working group is to come up with an engineering analysis, design and control protocol to address the "M1–M2" phenomenon in human disease. Therapeutic strategies that are incapable of distinguishing between these phenotypes and therefore destroy all immune function are suboptimal. Therefore, important goals include the minimally invasive characterization of the phenotype and the creation of therapies designed to shift the macrophage phenotype from M2 to M1—preserving the important role of the macrophages in the control and elimination of disease.

Elements of a possible approach include:

- Development of an overall strategy for the engineered design process.
- In addition to the overall strategy you will need a process of subdividing the design tasks and then integrating them. Consider the following required subsystems.
    o Sensors for the relevant biological signals
    o Logic processor for integration of input signals
    o Signal processing for a relevant set of actuators
    o Design and optimization of a control strategy
    o Consider fault detection and plan an abort module to cover unintended consequences
    o Integrate testing and validation of all subsystem models as the overall design progresses

An example that is already being developed (Anderson et al., 2005; 2007) is the engineering of a vehicle (bacterial) that can deal with the trade-offs of evading the host immune system long enough to be effective and delivering with specificity toxins to kill only tumor cells.

Required Reading

Anderson JC, Clarke EJ, Arkin AP, Voigt CA. Environmentally controlled invasion of cancer cells by engineered bacteria, *J Mol Biol* 2005;355:619-627.

DeNardo DG, Coussens LM. Balancing immune response: crosstalk between adaptive and innate immune cells during breast cancer progression. *Breast Cancer Research* 2007;9:212-222.

de Visser KE, Eichten A, Coussens LM. Paradoxical roles of the immune system during cancer development. *Nature Reviews Cancer* 2006;6:24-37.

Suggested Reading

Anderson JC, Voigt CA, Arkin AP. Environmental signal integration by a modular AND gate. *Mol Syst Biol* 2007;3:133.

Cold Spring Harbor Laboratory Proceedings from recent conferences and workshops on Engineering Design Principles in Biology. [Accessed online July 31, 2008: http://meetings.cshl.edu/meetings/engine08.shtm/engine08.shtml.]

Recent Conference on Synthetic Biology. [Accessed online July 31, 2008: http://syntheticbiology.org/ and http://sb4.biobricks.org/series/.]

TASK GROUP MEMBERS

- Amy Bauer, Los Alamos National Laboratory
- Chen Hou, Santa Fe Institute
- Wolfgang Losert, University of Maryland
- Roger Narayan, University of North Carolina at Chapel Hill
- Leor Weinberger, University of California, San Diego
- John P. Wikswo, Vanderbilt University
- Lani Wu, University of Texas Southwestern
- Mingjun Zhang, The University of Tennessee, Knoxville
- Hadley Leggett, University of California, Santa Cruz

TASK GROUP SUMMARY

*By Hadley Leggett, Graduate Science Writing Student,
University of California, Santa Cruz*

"Men at some time are masters of their fates. The fault, dear Brutus, is not in our stars, but in ourselves."

—*Cassius, from* Julius Caesar *by William Shakespeare*

When Shakespeare said that men are the masters of their own fates, he almost certainly wasn't talking about controlling destiny on a cellular level. But significant recent advances in the understanding of human biology have led scientists to wonder whether it might soon be possible to affect a person's future by controlling the fate of his or her individual cells.

By coaxing cells down a specific developmental path, scientists might be able to grow new pancreatic cells for a diabetic, or cure a patient's cancer by forcing wayward cells to stop dividing. The potential applications are endless, but the challenge of actually accomplishing this enticing goal is incredibly difficult.

At the 2008 meeting of the National Academies Keck *Futures Initiative* Conference, a Task Group (4) including biochemists, physicists and engineers met to wrestle with the enormity of the challenge.

"How do we cope with the huge size of networks that we find in biology?" asked Dr. Herbert Sauro, an associate professor of bioengineering at the University of Washington. "Estimates for different kinds of proteins in a single human cell may be up to 100,000," he said. "And who knows, the number of connections among those proteins could be much bigger."

To approach the problem, the group started with a specific example of regulating cell fate in the immune system of a cancer patient. Research has shown that the immune system plays a paradoxical role in cancer development: Some types of immune cells fight cancer, while others promote tumor growth. The group's task was to apply engineering-control models to create a cancer-fighting environment, rather than a tumor-promoting one.

How to Control a Macrophage

As described in the Challenge Summary to the group, immune cells called macrophages secrete specific signaling proteins depending on their phenotype. "Good" macrophages, called M1s, produce anti-tumor cytokines, while "bad" macrophages, called M2s, secrete growth factors that increase inflammation and encourage tumor growth.

Designing a macrophage control system to regulate cell phenotype means making a series of engineering decisions applicable to an active biological system. First, the group discussed general strategy. Should the focus be on changing M2 macrophages back into M1s, or preventing the development of M2 macrophages in the first place? Would simply eliminating all M2 macrophages create a tumor-fighting phenotype, or should the control system also generate more anti-tumor M1s?

A control system must also address specific engineering questions, such as whether to exert control from inside the cell—by altering gene expression, for example—or from outside, by changing the chemical landscape around the cell. Table 1 presents a list of specific design questions necessary for creating a macrophage control system.

Another important consideration is the need to anticipate unintended consequences. For instance, although M2 macrophages promote tumor growth, they also encourage wound healing and kill parasites. Some group members worried that eliminating M2 macrophages could slow tumor growth but create other problems, such as decubitus ulcers or chronic parasitic infections.

Essentially, the group decided the control system itself would need to be controlled. Instead of an open-loop system that operates without feedback, the system would need built-in monitoring and some kind of "volume knob," adjustable to maintain an ideal balance between M1 and M2 cells.

A Possible Solution: A Well-tailored Virus

One group member proposed a solution that could potentially encompass all of the design principles described in Table 1. Leor Weinberger, a virologist at the University of California, San Diego, suggested engineering a virus to infect fledgling immune cells and force them down the pathway of M1 development.

The virus would carry a set of genes that favored M1s and discouraged or destroyed M2s. It could exert internal control on infected cells by making them express M1 genes and secrete M1-friendly proteins. This would in turn exert external control on neighboring cells, by altering their chemical environment and encouraging them to become M1s as well. For instance, the virus might make infected cells secrete one protein to kill M2 macrophages and another to stimulate growth of M1s.

TABLE 1 Design Characteristics and Possible Solutions

Characteristics	Possible Solutions
Mechanism	Revert M2s to M1s
	Prevent M2 development
	Eliminate M2s
Application	Internal (e.g., altering gene expression)
	External (e.g., changing the microenvironment)
Range	Local (just around the tumor)
	Systemic (everywhere)
Timing	Continuous
	On-demand
Control	Open-loop (no control)
	Feed-forward (anticipatory control)
	Feed-back (outcome-based control)
Delivery method	Bacterial
	Viral
	Systemic injection with or without targeting

Best of all, Weinberger suggested, the virus could control expression of these genes by carrying an "inducible promoter"—essentially, a switch that turns on the genes only when there's an overabundance of M2s.

Unfortunately, using viruses to package and ship genes into cells is still difficult to do with the precision and safety needed for human medicine. To minimize risk, a virally delivered system would need to be carefully monitored and controlled.

In response to unintended side effects, one could keep adding more genes to the viral package so that each addition exerts a greater level of control over the system. However, a single virus can only fit a certain number of genes. "If you have a 64-kilobase plasmid, how many genes can you really control?" asked John Wikswo, a professor of biomedical engineering, molecular physiology and biophysics, and physics at Vanderbilt University.

By the second day of the conference, the group realized that many of their questions could only be answered through direct experimentation in the lab—either the tailored virus strategy would work, or it wouldn't. This is an experiment worth doing, and the group plans to explore this strategy after the conference.

Creating a General Model

For the rest of the weekend, the group focused on applying engineering principles to controlling cell fate in a general context, not specific to immunology. Like any good control system, a system to control cell fate would require sensors, actuators and mathematical models. The sensors would need to detect what's going on both inside and outside of the cell, and they might take advantage of recent advances in silicon technology. The actuators would do the real "work" of the system—in other words, they would take information provided by the sensors and translate it into whatever action would stabilize the system.

For instance, in the example of macrophage phenotypes, the sensors might detect an overabundance of M2 cells, and the actuators (in this case, a virus) would respond by turning on genes to correct the balance between M1 and M2 cells.

But to know what action to take in a given situation, a system to regulate the phenotype of a cell would need accurate mathematical models describing cell behavior. Creating these models could present a huge hurdle: Given the hundreds of thousands of proteins in a cell, it would be extraordinarily difficult to accurately describe every interaction. For the present, the group recommended a "black box" model that would deal only with the level of detail necessary to achieve a given outcome. Basically, learn only what you need to know, and dump everything else in a black box.

Creating that kind of model would require what engineers call a "system decomposition"—essentially, breaking down cellular function into discrete modules and identifying specific inputs and outputs.

Once again, this task requires venturing back into the lab, where researchers could perform many experiments in parallel, making small adjustments to see which changes made a difference in the stability of the system.

In Task Group Four's final presentation, Wikswo from Vanderbilt left conference participants with a final, somewhat disturbing thought: In the end, it's quite possible that systems to control cell fate might turn out to be just as complex as the organisms they're meant to control. He compared controlling cell fate to trying to fly the X-27 fighter jet, a highly complex experimental plane that never made it past the mock-up phase. "You might have to turn a lot of knobs at once," he said, "to keep the system in the air." The biological challenge is to identify strategies that control the system on a coarse-grained scale and don't require turning the 100,000 or more knobs within each cell. Obviously many single-target drugs can control some cells. What's needed now is to address distributed, multitarget cell sensing and control.

Task Group Summary 5
How can social networks aid our understanding of complexity?

CHALLENGE SUMMARY

Society represents one of the most complex systems that science has ever encountered. Its six billion individuals interact with varying frequencies, generating a complex social network that plays a key role in the spread of ideas and political systems, the breakout of riots and wars, and the health and well being of most individuals. Despite the obvious scientific and economic importance of understanding social networks, today we know more about *E. coli* bacteria than about patterns of human interaction, partly because bacteria do not get annoyed when we put them under the microscope. Most of the research on social networks focuses on snapshots of small scale networks; there is a general paucity of research on the dynamics of large scale human interaction. Our lack of understanding of macro social networks and human behavior is not due to the lack of technologies to collect the relevant data: thanks to the computerization of most aspects of life, today an increasing amount of information is automatically collected about all of us. Mobile phone companies know who calls whom and where their consumers are; email providers keep detailed records of their consumers' electronic communications; credit card companies and banks can piece together not only their consumers' wealth and spending patterns, but also their travel patterns. Taken together, society is the only complex system whose components are constantly monitored, offering a potential testing ground for all complex systems theories of quantitative predictive power. However, a major problem is that most of the collected data are owned by private organizations and protected by layers of confidentiality agreements and economic interests that keep scientists at arm's reach from making use of them. (Ideas about privacy and the technologies that protect it are a topic in themselves.) While for decades complexity was driven by theoretical ideas, the current availability of large quantities of data is rapidly turning complexity into an empirical science, offering significant opportunities for complexity to show its relevance to society.

The lack of access to relevant datasets demonstrates that the scientific community has failed to explain the scientific and societal benefits of understanding human behavior. The average person understands the need to invest in biomedical research, expecting that the results will lead to better approaches to the prevention or treatment of disease. Most people also comprehend the need to study materials science, which may result in better computers and phones. Yet society in general does not recognize the value of research on data that are already collected on human behavior because there is no ready understanding of its potential value but clear concern about invasion of our privacy. Credit card data are an example. If researchers could collect anonymous data, that is without names or credit card numbers but with data about geographic location, age, and ethnicity it would be possible to develop significant understanding of buying habits and purchasing patterns. For scientists to have access to these datasets, they will need to present a compelling explanation of the scientific benefits of these datasets. Much of the discourse in the scientific community has focused on the problems surrounding user confidentiality and the adverse effects of such data collection processes, rather than the benefits of a systematic program in social complexity. Consequently, investment in quantitative social sciences represents a tiny fraction of the current federal research budget and it is virtually not existent in the industry.

Key Questions

The challenge to this working group is identify one or a few research areas that would clearly illustrate the societal benefits of getting access to the currently collected high quality datasets on patterns of human activity, spelling out the societal benefits of this research in a fashion that will be obvious to taxpayers, decision-makers and database owners alike.

- Identify one or several key questions pertaining to social networks and human behavior that are of fundamental importance for complexity science and offer significant potential societal payoffs.

- Discuss how such a proposal should be presented to the society and decision-makers to address and manage the privacy needs of individuals, or to convince these constituencies that the benefits outweigh the risks involved in such research.
- Identify the type of data necessary to address the proposed question, and the private or government sources that either own or have access to useful datasets.
- Identify the magnitude of the investment and the ideal mechanisms necessary to make the research program a reality.
- Explore to what degree such a program would answer questions that benefit not only the understanding of social systems, but also uncover laws and mechanisms that other systems of comparable organization and complexity.

Required Reading

Butler D. Data sharing threatens privacy. *Nature* 2007;449:644-645.

Moody J. The importance of relationship timing for diffusion: indirect connectivity and STD infection risk. *Social Forces* 2002;81:25-56.

Onnela J-P, et al. Structure and tie strengths in mobile communication networks. *Proc Natl Acad Sci USA* 2007;104(18):7332-336).

Duncan J. Watts. A twenty-first century science, *Nature* 2007;445-489. [Accessed online August 1, 2008: [http://www.nature.com/nature/journal/v445/n7127/full/445489a.html].]

Suggested Reading

Palla G, Barabási A-L, Vicsek T. Quantifying social group evolution. *Nature* 2007;446:664-667.

Putnam R. Bowling alone: America's declining social capital. [Accessed online June24,2008: http://xroads.virginia.edu/~hyper/DETOC/assoc/bowling.html.] [Published online on April 24, 2007, 10.1073/pnas.0610245104 2007.]

Stark D, Vedres B. Social times of network spaces: Network sequences and foreign investment in Hungary. *American Journal of Sociology* 2006;111(5):1368-1411.

TASK GROUP MEMBERS

- Timothy B. Buchman, Washington University in St. Louis
- Kee Chan, Boston University
- Noshir S. Contractor, Northwestern University
- Nathan Eagle, MIT/Santa Fe Institute

- Joshua Epstein, The Brookings Institution
- Robert A. Greenes, Arizona State University
- Alan J. Hurd, Los Alamos National Laboratory
- Cristopher Moore, University of New Mexico
- Joshua Plotkin, University of Pennsylvania
- Richard Puddy, Centers for Disease Control and Prevention
- Jordan Sarver, University of Georgia

TASK GROUP SUMMARY

By Jordan Sarver, Graduate Science Writing Student, University of Georgia

The human body is full of complex systems. The nervous system, the circulatory system, and even the immune system consist of a series of smaller components working together in the body. For years, scientists have been able to view many of the networks that exist inside the body. On a macro-scale, human relationships are an example of smaller components working together to produce a larger result as well. Disease epidemics, religions, and even worldviews are spread through human interaction. Unlike body systems, there is no tool capable of viewing all of the connections that a person makes over time; nor is there a method known to follow the propagation of an idea throughout a community. At the 2008 meeting of the National Academies Keck *Futures Initiative* Conference on Complex Systems, a multidisciplinary group of researchers focused their energy on the complex system of social networks.

The personal connections that people create and destroy everyday form what are known as social networks which influence every aspect of our daily lives from health to knowledge. When someone makes a new friend their network increases. A new friend provides access not only to new information, but also other contacts as they meet people within someone else's network. An example of how social networks can affect a person's health is the flu. Imagine a passenger on an airplane with the flu virus—everyone on the airplane is exposed to this virus as well as anyone who comes into contact with the passenger throughout his travels. The spread of the flu virus is a direct result of the contact the carrier has with other people throughout his day.

The group set out to find a way to interpret and map the networks that exist between people, places, and ideas. First, the group wanted to figure out a way to take the information that is already public and available and then construct a network connecting people to places and beliefs. Some in the group believed the problem is that scientists have not sifted through current information properly and acquiring more would only complicate current advancements. Another problem encountered when studying social networks is the dynamic nature of social systems. People are constantly making and losing connections with other people and ideologies. Constant changes realign the threads that create social networks and makes mapping the network that much harder.

Technology has become an ally in the efforts to map social networks. The Internet has become an optimum tool for monitoring human behavior. Websites like Facebook and MySpace track whom people befriend and to which organizations they belong. Accordingly group members discussed the possibility of being able to map connections and the relationships that exist between people in society using the web as a tool. Social websites provide an easier way to look at some complex systems that are formed all the time. These websites could potentially be used as a tool to track the path of information through cyberspace.

Topics and questions that tracked the flow of conversation included how information is disseminated throughout networks (Internet, email), why some people are superspreaders of information or disease, what network structures are required for optimization, how innovations are diffused over the Internet, and what role team science plays in social networking.

Not all information is easily accessible. Americans are extremely private people. Media has allowed Americans to watch, read, and listen to things that interest them without being subject to public scrutiny. In turn, an environment where privacy is extremely valued has been created. Several types of information that are currently protected by privacy laws could potentially reveal helpful information. Medical records are kept in strict confidence by healthcare professionals, but group members noted the value of linking the spread of disease with medical records. Constructing a network of disease transmission could not only lead to more effective treatment, but it could also provide information that will allow for preemptive measures to halt the spread of certain diseases.

Many of the group members brought along their notebook computers and used various websites as sources of information about social networks. The alleged link between autism and childhood vaccinations was a case study used by the group to look at how a specific idea flows through cyberspace. Using Google Trends, a subset of the Google search engine, group members were able to see how many Internet users searched the Internet for a link between autism and vaccinations. The website provided specific information about which states searched for the topic of autism and vaccinations the most and which websites were the most viewed on the subject.

Information from the Centers for Disease Control and Prevention along with the Google Trends data allowed the group members to construct a preliminary relationship between Internet searches and real-world effects. Group members examined data from Iowa where there was a mumps outbreak in 2006 and the relationship, if any, to the belief that childhood vaccinations cause autism. The lack of vaccinations appeared to correspond with the outbreak, but there was no concrete evidence available to prove that the lack of vaccinations was a result of the belief that vaccinations cause autism. The ability to track the spread of opinion with the spread of disease could potentially allow those in the field of prevention to be more specific in targeting at risk populations.

Another example included the recent rise and spread of "direct-to-consumer genetic testing" as a potential topic to explore further. Both text-mining and web crawling were highlighted as potential techniques to explore multi-dimensional networks using such sites as the "thewaybackmachine," "Digg," "Xobni," "dotnetmap," "Google Trends," "Facebook," etc.

Another important aspect of social networks is the idea of team science. Members of the group were intrigued by the thought of creating a way to predict likely collaborations among scientists across disciplines. In fact, it was noted by one of the group members that articles with the highest impact include multiple authors across several disciplines. An article will have an even higher impact if those authors from different disciplines are geographically diverse. What was developed in the group was the idea of creating a way to predict collaboration using subject matter of an author's previous work, proximity to other scientists, and previous collaborators.

Naturally, group members decided that more information is needed. Medical records are highly privatized, but full of rich information, and so the issue is whether scientists can encourage people to be more open with their private information in the interest of creating large, medically useful databases for the common good. An equally important goal is to encourage the exchange of information among health professionals who have a need to know.

What are the big picture ideas? Constructing a way to facilitate team science across a multi-disciplinary field was a major priority of this group. These group members from different disciplines are an example of the extraordinary results that can be reached when different scientific minds collaborate. Mapping the propagation of an idea or an action with the propagation of a disease is now possible using the information available. Also, human behavior influences disease trends which in turn, induce more human behavior. The group members realized the capabilities of the Internet to take snapshots of trends in both cyberspace and the real-world. By assembling these snapshots, scientists can produce a moving picture and watch the progression of either disease or ideologies across time and space.

The next step for the group was developing a way to construct a network and watch it grow. Facebook, the social website designed for college students, seemed like an optimum way to watch the creation and spread of a network. Various applications exist on the Facebook website. These applications allow members of the website to play games or connect with friends through shared interests. The group proposed tracing the application that allows members to either become a vampire or a werewolf. Members can then bite friends, making them a member of either the vampire or werewolf group. Although the application is fairly innocuous, the information could be useful. An application like the one on Facebook could predict trends. Members of a group who are influential and the conditions that predict who will become a part of a new network could all be revealed using the simple application. The results could be wide reaching in terms of predicting how information will spread through cyberspace. It could also predict how disease will progress by determining who is likely to be a super-spreader, someone who interacts with numerous amounts of people. By identifying super-spreaders and who they will potential interact with, prevention could become more efficient.

The Internet is both a source of information and a tool to decipher information. Social science has potentially reached the end of a paradigm that has existed for a long time—the lack of understanding of contact patterns of populations. Now there is data to address the challenges that the field has faced. Flow can now be visualized. Scientists are no longer relegated to watching flow lines and calculating equations. The capacity to examine minute by minute construction of our world through relationships is now possible.

Other research priorities include: computational thinking, making the implicit explicit, capturing relational data on a massive scale, accounting for visualizations and dynamics in this new environment, moving society from a snapshot to a moving picture of social networks, and investigating how ideas spread as well as how to spread ideas.

Task Group Summary 6

The brain is the epitome of complexity. How will understanding the complex, linked interactions among the many types of neurons in the brain lead to knowing how the brain contributes to normal function and susceptibility to neuropsychiatric disease?

CHALLENGE SUMMARY

The human brain, especially our cerebral cortex, is responsible for the sophisticated thoughts, memories, perceptions, and language that distinguish our species from all others. These functional abilities are the result of a complex, prolonged developmental history that involves expression of about half of the genes in our genome and proliferation, migration, and differentiation of scores of different cell types. This is especially evident in the human cerebral cortex, a multilayered structure that is roughly 3 times larger than that of our nearest primate ancestors. Correspondingly, molecular analysis suggests that these human-specific characteristics are associated with accelerated rates of evolution of the protein products of the genes implicated in the development of the human central nervous system that are higher in primates than in other organisms

These complex developmental programs and processes not only are responsible for the enhanced functional abilities of the human brain but are also error prone and likely to contribute to common complex disorders of the central nervous system (CNS) such as schizophrenia, bipolar disease and obsessive-compulsive disorder, conditions that in aggregate affect 2-3% of adults. Understanding the etiology of these multi-factorial diseases, each of which appears to be the result of both genetic and environmental variables, and developing effective strategies for their treatment and/or prevention is a major contemporary challenge for medicine and biomedical research.

Key Questions

- What are the evolutionary forces driving the rapid evolution of the human brain and what are their consequences for the sources and frequencies of neuropsychiatric disease?
- Can genetics and genomics identify all the genes involved in the development and function of the central nervous system?
- Can we understand how the protein products of these genes integrate into biological systems essential for CNS development and function?
- What are the components, structure, and behavior of the biological systems that underlie complex CNS functions such as memory, reasoning, and language?
- How do combinations of variants in a subset of these genes and proteins perturb the function of the biological systems characteristic of the CNS and increase risk for neuropsychiatric disease?
- What technologies and resources, existing and yet to be developed, would improve our abilities to understand normal and abnormal brain development and function?

Required Reading

Bystron I, Blakemore C, Rakic P. Development of the human cerebral cortex: Boulder committee re-visited. *Nature Rev NeuroSci* 2008;9:111.

Dorus S, Vallender EJ, Evans PD, Anderson JR, Gilbert SL, Mahowald M, Wyckoff J, Malcom C, Lahn BT. Accelerated evolution of the nervous system genes in the origin of *Homo sapiens. CELL* 2004;119:1027.

Hill RS, Walsh CA. Molecular insights into human brain evolution. *Nature* 2005;437:64.

Pollard KS, Salama SR, Lambert N, Lambot M-A, Coppens S, Pedersen JS, Katzman S, King B, Onodera C, Siepel A, Kern AD, Dehay C, Igel H, Ares M, Vanderhaeghen P, Haussler D. An RNA gene expressed during cortical development evolved rapidly in humans. *Nature* 2006;443:167-172.

Sawa A, Snyder SH. Schizophrenia: Diverse approaches to a complex disease. *Science* 2002;296:692-695.

Ross CA, Margolis RL, Reading S, Pletnikov M, Coyle JT. The neurobiology of Schizophrenia. *Neuron* 2006;52:139-153.

Due to the popularity of this topic, two groups explored this subject. Please be sure to review the second write-up, which immediately follows this one.

TASK GROUP MEMBERS – GROUP A

- John M. Beggs, Indiana University
- Sally Blower, David Geffen School of Medicine at UCLA
- Stephen J. Bonasera, University of California, San Francisco
- Nick Ellis, University of Michigan
- Veit Elser, Cornell University
- Daniel Fletcher, University of California, Berkeley
- Philip LeDuc, Carnegie Mellon University
- Andreas Trabesinger, *Nature Physics*
- Shyni Varghese, University of California, San Diego
- Larry Yaeger, Indiana University
- Lizzie Buchen, University of California, Santa Cruz

TASK GROUP SUMMARY – GROUP A

By Lizzie Buchen, Graduate Science Writing Student,
University of California, Santa Cruz

The Problem

The philosopher John Stuart Mill once marveled at the combustion of methane: What went in—a violently flammable fuel—bore no resemblance to what came out—innocuous water and carbon dioxide. The scientific understanding of the 1800s could not account for this seemingly miraculous transformation.

Likewise, most people are stupefied when pressed to explain the mind: What goes in—the electrical and chemical interactions of 100 billion cells, agglomerated into three pounds of fatty flesh—seems to have no relation to the phenomena that emerge—emotions, imagination, abstract reasoning, physical dexterity.

Today, the subject of Mill's wonder is far less mysterious; developments in chemistry and physics explain chemical reactions as the predictable movements of electrons between atoms.

Neuroscientists hope for a similar outcome—that a more thorough comprehension of the brain's components and their interactions will explain its remarkable output.

In recent years, our understanding of the brain has burgeoned. We are learning how currents flash through neurons, how neurons are born and how they die, how connections between them develop, strengthen, and fade away, and how different regions of the brain interact. We can even replicate small portions of the brain *in silico*—as IBM's Blue Gene supercomputer did with a cubic millimeter of the cortex in 2006. And yet, we seem no closer to understanding how all this neuronal chattering manifests as the mind. It is clear that an approach that might work for the brain will be different from the reductionist one, where general properties of the grand structure are indispensible.

At the 2008 National Academies Keck *Futures Initiative* Conference on Complex Systems, a group of scientists charged with designing a protocol for 'the brain as the epitome of complexity,' decided to develop a strategy for teasing out the principles that govern this metamorphosis. The team was composed of engineers, neuroscientists, epidemiologists, computer scientists, physicists and psychologists.

Neural Complexity

The group viewed the brain as a quintessential complex system: it consists of fairly simple components (neurons) that engage in coordinated interactions, which are somehow bound or integrated to produce complex emergent phenomena (thoughts). "Complexity" in the brain refers to the structure and behavior of these interactions—the physical connections traveling forward, backward, and laterally between various regions of the brain, as well as the timing of the communications.

"If you have all your neurons firing randomly, with only short-range connections, that's not very complex," Larry Yaeger of Indiana University observed. "But if you have them all synchronized, all firing in lockstep, that's not complex either. The good stuff is in the middle."

"The good stuff"—a highly complex brain—has a balance between these extremes of organization: neurons that are coordinated mostly with their close neighbors, but also communicate with other neuronal neighborhoods. A brain with specialized but interconnected regions—such as a region that processes vision connected to a region that generates movements—is necessary for complex behaviors, like stopping at a red light.

Complexity, in this formal sense, is a way to quantify how the brain is organized, and so is directly correlated with how the brain works. The group thinks differences in neural complexity is likely to account for differences in intelligence. For example, animals capable of abstract reasoning will exhibit greater complexity than less intellectually capable animals. By dissecting this complexity measurement, one can understand what organizational principles make the region complex—how many connections the neurons make, what type of connections they make, how each neuron behaves. This may enlighten understanding of how the brain achieves abstract reasoning.

An important control when measuring complexity is size—an elephant's brain has about four times as many neurons as a human's, yet we assume it is less complex. Although more neurons may result in more connections and potentially more behaviors, the connections may be irrelevant or even detrimental to functioning. It is the organization—not sheer number—of the brain's connections that result in intelligence; complexity captures this organization.

The Approach: Focus on Impulse Control

To use neural complexity as a probe for understanding how the brain produces intelligence, the group found it helpful to focus on a microcosm of intelligent behavior: impulse control. The human ability to voluntarily postpone gratification for the sake of later outcomes vastly exceeds that observed elsewhere in the animal kingdom—humans have the ability to abstain from drugs and sex, they diet, they save money, some even go to college and professional schools.

Primates, too, can delay gratification, picking large delayed rewards over smaller immediate rewards—but only if the delay is on the order of minutes. Mice can only delay gratification for a few seconds, and this ability differs between different strains.

A chief reason for selecting impulse control is its relevance to psychiatric disease. People with schizophrenia, bipolar disorder, and obsessive compulsive disorder have poor impulse control—as do "normal" individuals if they have had a bit too much to drink.

Impulse control is also related to more subtle differences in human behavior. A striking 1989 study demonstrated that impulse control in 4-year-olds is predictive of their later success in life.

In the experiment, a researcher placed a marshmallow in front of a child and told him he would return in 20 minutes with a second marshmallow—but would only give it to the child if he had not eaten the first before the researcher's return.

The study showed that children who delayed gratification and waited for the second marshmallow developed into more cognitively and socially competent adolescents. They were more likely to go to college, less likely to be arrested, and less likely to develop eating disorders. The measure of impulsivity was more predictive of their success in life than IQ.

Impulse control, then, is a specific, measurable behavior that is relevant to intelligence, making it ideal for probing the relevance of neural complexity. The group hypothesized that specific aspects of the brain's organization, quantified as complexity, will be predictive of impulse control.

The Plan of Attack

1. For a detailed measurement of neural complexity, it is essential to gather data: the connections, communications, and firing patterns of as many neurons as possible. This requires great advancements in technology—implanting tens of thousands of recording electrodes, for example, and imaging anatomy with much improved resolution in both space and time. In high numbers, these data points would provide insight into higher levels of cognitive processing.

2. The group proposes to process the data by calculating neural complexity. There are a number of equations and models that quantify complexity, each looking at different aspects of the brain's organization—timing of communications, number and type of connections, etc.

3. The group would compare these different complexity measurements between organisms with different abilities to control their impulses—different strains of mice, different species, humans with certain diseases, and humans with different skills, such as artists and scientists. This will reveal which aspects of the brain's organization are related to impulse control.

If this strategy is effective, the group will apply it to other intelligent behaviors, such as language. The group hopes to understand which aspects of the brain's organization are linked with intelligent behaviors—developing a complexity "signature." This knowledge will enlighten our understanding of the relationship between the brain's complicated form and phenomenal function.

Applications

A key ambition of the group is to use its strategy to benefit society. The presumption is that understanding impulse control is important to many psychiatric diseases—not only for diagnosis but also for therapy. For example, if it is possible to use measures of complexity to pinpoint the precise mis-wiring or mis-firing that leads to impulsivity in obsessive-compulsive disorder, it might be possible to use this measure to evaluate potential interventions.

The brain is a daunting enigma for today's neuroscientists, but the group is confident that an understanding of its underlying principles is in the foreseeable future. The inevitable advances in neuroscience technology will give researchers a real-time view of neuronal interactions across the entire brain; by analyzing the complexity of these interactions, the group hopes to unravel how the phenomenal mind emerges from the physical brain.

TASK GROUP MEMBERS – GROUP B

- Craig Atwood, UW – Madison
- Edward Boyden III, MIT
- Tansu Celikel, University of Southern California
- Eugenio Culurciello, Yale University
- Rhonda Dzakpasu, Georgetown University
- Sarah Heilshorn, Stanford University
- Christopher Kello, University of California, Merced
- Daniel Lathrop, University of Maryland
- Brian Litt, University of Pennsylvania
- Stefan Maas, Lehigh University
- Olaf Sporns, Indiana University
- Dagmar Sternad, Northeastern University
- Jennifer Lauren Lee, University of Southern California

TASK GROUP SUMMARY – GROUP B

By Jennifer Lauren Lee, Graduate Science Writing Student,
University of Southern California

Every year, new and more sophisticated methods of investigation bring the workings of the human brain into sharper relief. Yet the more details we gather, the less clear it is where the journey to a complete understanding of the brain will end; each new rise in knowledge reveals a horizon still out of reach. The brain is composed of complex systems (cells) with highly diverse and plastic connections that distinguishes it, and in turn its properties, from many other types of complex networks. A full understanding of the brain could provide innumerable boons to the field of medicine, granting physicians the ability to diagnose neurological diseases more quickly and treat them more effectively. At the 2008 meeting of the National Academies Keck *Futures Initiative* Conference on Complex Systems, one multidisciplinary Task Group (6B) was determined to see whether treating the brain as a *complex* system might spark ideas for new tools to help scientists understand the brain as a *complete* system.

The Opportunities of Neuroscience

Early in the discussions, the members of this task group were concerned with the problem of scale. Each year brings improvements in the techniques that allow scientists to probe the brain at many levels—that of protein structure, for example, or single neurons interacting with one another, or entire sections of the brain that each consist of millions of neurons working together as a unit. But what these technological improvements do not do is improve scientists' ability to see how the various levels connect with one another. The "rules" for neuron-to-neuron interaction, as compared to those governing the relationship between two zones or areas in the brain, for example, are so different that a person can spend an entire career studying a single level of interactions without ever looking beyond. In a sense, each scale in the brain is a separate field of study, with its own jargon and techniques for collecting data—an island in the ocean of brain science.

These gaps between the scales are unknown territories in studies of the brain—what one member of the group called the "wastelands of neuroscience." And it was these lacunae that became this group's focus.

One of the first orders of business was defining terms, so that researchers with different areas of expertise could be sure their words meant the same thing to everyone at the table. The brain is always active—"till you're dead," as one participant put it. But it can exhibit what could be called different "states" depending on what it is doing. Taking a snapshot of the complete activity on every scale of the brain in a given state would yield what could be called a "signature" for that state. A healthy brain would have the healthy signature for juggling, or sleeping, or looking at the color blue, while doing each of those tasks. A diseased brain—one with the earliest signs of epilepsy or Alzheimer's disease, for example—might have an abnormal signature; its pattern of activity for a given task would be different, in theory, on at least one scale. The size of the difference would determine when and how the disease manifests itself, and how quickly it progresses.

The team also considered the possibility that neurological diseases might affect the level of complexity itself, possibly lowering the brain's complexity and reducing its ability to respond to problems. The challenge, then, would be to make a model that shows the relationship between the various scales, using the tools of complexity to analyze data at each level simultaneously. In this way, one could determine the characteristic "disease state" for a particular activity.

Brave New Methods

In order to "see" the connections between the scales, the group decided it would need to study various levels of the brain *at the same time* in response to some stimulus. Getting a sense of how the various levels interact with one another would give the team a signature for that particular brain state. The first step would be finding the complexity signature of the resting state of a healthy brain. Then researchers would perturb the system, and see how those perturbations affected the other scales. They could make changes at the smallest scale—that of genes and proteins—then track those changes through the higher levels, up through the largest networks of neurons in the brain. They could also use a top-down approach, perturbing the whole system (through sleep deprivation or a behavioral change, for example) and observing what happens at the smaller scales. Researchers would start by using existing techniques, such as probing individual neurons with fluorescent imaging or assessing the activity of larger areas with functional magnetic resonance imaging (fMRI). But the scientific community would also need to develop "Brave New Methods," new tools to "see" changes at each scale and map those changes together.

Also necessary would be a method of connecting the different scales, to catch the changes in the brain's activity signature at each level in response to the task being performed. Here the group ran into some hypothetical problems. How would they know whether they had matched up the scales correctly, given the different methods (each with its own types of errors) they had used to collect the information at each scale? How would they decide how many scales to consider, and how to break them up? And how could they know when they were finally looking at a complete system—that, as one task group member put it, they had the whole system in their scopes?

Without brave new methods, the immediate answer would be to collect data—a lot of data—and compare the results with models that would reconstruct the missing points between the layers in space and time. The only way to validate a model is to test how well it predicts the results of the next data collection. The more data, the more sophisticated (and, presumably, reliable) the models.

Waiting for Symptoms

Although the techniques for conducting this research need to be refined, the benefits could revolutionize humanity's understanding of the brain and also the facility with which brain diseases such as epilepsy are treated. It could take ten years after an injury for the first symptoms of epilepsy to present themselves as a seizure; and by then, perhaps, the damage is done. If measuring the changes in the complexity of the system could allow scientists to catch the earliest signs of a disease, regardless of the scale on which it presents itself, patients might have a better chance of recovery.

This new way of mapping the brain using complexity may also provide researchers with a short-cut to a functional understanding of the brain. One member of the group compared the practice of studying the brain on a neuron-to-neuron level to that of trying to understand the economy by following all the shoppers in a supermarket: although these details may give the viewer insight into one level of the system, they do not give much useful information about the system as a whole. A method of studying the brain that makes use of complexity theory might allow us to get a full picture of how the brain "works" before we have finished defining the roles of every gene and protein in the body. With any luck, this new view could yield incalculable benefits to medicine. In the meantime, it would provide a brave new way of thinking about the brain—a way that might inspire people to create new models and tools for tackling a new problem.

Task Group Summary 7
How can we enhance robustness of engineered systems, and how can the methods of engineering analysis be extended to address issues of complexity and management in other fields?

CHALLENGE SUMMARY

The support of policy, industry, or private decisions involving complex, dynamic systems and uncertainty, is a challenge that presents common features across different fields. For example, lifecycle risk management in the automotive, space, and medical device industries involves complex physical systems, organizations, and uncertainties that vary with experience (test results, operational data, etc.). Similarly, the maintenance of the heat shield of the US space shuttle involves the physical characteristics of the tiles as well as human and organizational factors (including errors). The methods of engineering analysis can be extended beyond the realm of engineered systems to address issues of complexity and management in other fields.

For instance, the design and operation of health care systems include both technological and human factors: how can information and incentives best be managed to enable affordable, quality healthcare, given the complex hierarchical domains involved, with levels ranging from clinical practices to the delivery of care and specific organizations? Another example is the management of the Internet, whose structure and interactions with different markets evolve constantly, requiring an understanding of both the network and the complex behaviors of their users.

One common thread is the engineering approach that can be adopted for the design and management of such complex systems, with an emphasis on architecture (structure and functions) and a systematic, coherent treatment of both dynamics and uncertainties. One of the challenges is to build in and preserve robustness and adaptability, accounting for complex interactions among components; for example, to include interfaces and interactions of systems with the medium in which they operate and to anticipate future performance *in situ* (the human body for medical devices, soil/structure interactions, variations of external parameters of space for satellites, etc.). At another level, these interactions include those between the physical system and its operators (pilots, technicians, doctors), and between these operators and the managers who set the incentives and the information base for the people in charge of operations. The goal is to adapt the methods of engineering systems analysis to other types of complex systems (human, climatic, etc.) in order to support policy decisions before full information has been gathered.

Decisions pertaining to the management of design, tests, development and operations can be supported by a combination of systems analysis (static and dynamic), risk analysis and decision analysis. In addition, methods of economic analysis (including for instance, utility theory, principal-agent models and game theory) allow us to evaluate questions such as incentives as well as issues about budget optimization.

Uncertainties are often at the core of the problem. In the context of risk management, one can rely on classical statistics when that information exists and the system is stable enough; but these data are not always available or relevant to all challenges—for instance in the design stage of new devices. Bayesian probabilities are useful to support risk management decisions, in all phases of a device life (design, testing, approval, operation, and retirement). The challenge is to combine the powers of all existing methods to make the best possible use of incomplete information in the management of complex systems, both in industry and in government.

Key Questions

The key question is: how can the methods of engineering analysis of complex systems be extended to other types of systems (human, biological, physical), medical systems (e.g., anesthesia in operating rooms), threats of terrorist attacks, climatic phenomena, etc.? Problems can arise at the interface of engineered systems and the medium in which they operate, or the organizations that manage them. These interactions and the corresponding uncertainties have to be accounted for in a systematic way to support rational decision making. One focus can be the assessment and management of the risks of system failures and/or of reduced levels of performance based on concepts of systems analysis, probability, stochastic processes, and economic analysis.

The challenge to the working group is to come up with engineering strategies to address the fundamental problems of information and decision-making associated with the management of complex systems.

Required Reading

Carlson JN and Doyle J. Complexity and robustness. *Proc Natl Acad Sci USA* 2002;1:2538-2545.

Overview of the Vatican workshop of 1999. [Accessed online June 10, 2008: http://www.vatican.va/roman_curia/pontifical_academies/acdscien/documents/rc_pa_acdscien_doc_20000530_survival_en.html.]

Paté-Cornell ME. The engineering risk assessment method and some applications. In: W. Edwards, R. Miles, and D. von Winterfeldt (eds.), Advances in decision analysis. New York: Cambridge University Press 2007.

Suggested Reading

Basole RC, Rouse W. Complexity of service value networks: conceptualization and empirical investigation. *Systems Journal* 2008;47(1):53-70.

Davis JP, Eisenhardt KM, and Bingham CB. Complexity theory, market dynamism and the strategy of simple rules. Stanford University, Department of Management Science and Engineering, 2007. [Accessed online July 31, 2008:http://web.mit.edu/~jasond/www/complexity.htm.]

Murphy DM and Paté-Cornell ME. The SAM framework: a systems analysis approach to modeling the effects of management on human behavior in risk analysis. *Risk Analysis* 1996;16(4):501-515.

Paté-Cornell ME and Fischbeck PS. Probabilistic risk analysis and risk-based priority scale for the tiles of the space shuttle. *Reliability Engineering and System Safety* 1993;40(3):221-238.

Rouse W. Complex engineered, organizational and natural systems: Issues underlying the complexity of systems and fundamental research needed to address these issues. *Systems Engineering.* 2007;10(3):260-271.

TASK GROUP MEMBERS

- Fahmida Chowdhury, National Science Foundation
- Jeffrey Cooper, SAIC
- Tuan Duong, JPL/CIT
- Theirry Emonet, Yale University
- James Ferrell, Stanford University

- Panos Papadopoulos, University of California, Berkeley
- Michael Leo Parchman, South Texas Veterans Health Care Systems
- Vimla L. Patel, Arizona State University
- Steven Schiff, Penn State University
- Jessika Trancik, Santa Fe Institute
- Andreas Wagner, University of Zurich
- Joseph Wang, Virginia Polytechnic Institute and State University
- Cassandra Brooks, University of California, Santa Cruz

TASK GROUP SUMMARY

*By Cassandra Brooks, Graduate Science Writing Student,
University of California, Santa Cruz*

Human beings have long used engineering principles to solve complex problems, but these systems aren't infallible and increasing their robustness is a pressing concern.

With this theme in mind, 11 scientists from different engineering and biological fields met at the 2008 National Academies Keck *Futures Initiative* Conference on Complex Systems to discuss their assigned question: How can we enhance robustness of engineered systems, and how can the methods of engineering analysis be extended to address issues of complexity and management in other fields?

Robustness refers to the ability of a system to preserve itself in response to perturbations. In other words, a robust system is one that can withstand variations with minimal damage or loss of function. Examples are buildings designed to maintain their integrity during an earthquake, power cords with built in surge protectors, and a mammal's ability to maintain a constant internal temperature in different climes.

Specific characteristics generate robustness in a system: redundancy, control systems, distributed robustness, error-correction and hardness. Redundancy is the duplication of critical components that will increase the reliability of a system. Control systems are devices that manage or regulate the system to keep it functioning properly. Distributed robustness means the robustness is spread throughout the system. Any system will fail at its weakest point. Error-correcting systems simply refer to the system's ability to detect and fix errors without perpetuating them. Lastly, hardness means over-designing something to make it stronger. For example, an aircraft that has two engines (with one for back up) is redundant, whereas a bridge designed to remain standing in winds exceeding what engineers expect to see in nature is hardness.

As the Task Group (7) discussed various specific engineering fields and broader aspects of biological systems, an underlying theme arose. Biological systems are inherently robust. Gene flow, genetic drift, natural selection, non-random mating, and mutation (the five mechanisms of evolution) result in the most robust of systems because with living organisms, health and proper function must be the norm.

The group generated questions spanning biology and engineering and clustered them according to common themes. Why are most engineered systems rigid while biologic systems are soft? Which engineering principles for robustness are applicable to human/social systems? And which engineering principles are not found in biologic systems and vice versa?

The group focused on the latter part of the last question, "Which biological systems are not used in engineering?" to address its ultimate question, "What complex biological behaviors or systems can be applied to solve engineering problems and make engineering systems more robust?"

Uncertainty and human error are major problems compromising the robustness of engineering systems. An example would be the maintenance of the heat shield on United States space shuttles, which requires precise engineering as well as human and organizational factors. As we saw in 1993, human error and organizational problems at NASA led to the devastating Challenger explosion. The group began to ponder: Can we engineer a system to adapt and regenerate despite perturbations caused by human error or other uncertainties?

Consider regeneration from a biological perspective. Regeneration, or the replacement of a defective limb, is a terrific example of redundancy in nature. Imagine if engineered systems could adapt to a problem by spontaneously fixing themselves. What if we could engineer a space ship to regenerate broken parts? What if we could somehow manufacture cells that would replace the damaged heat shield in the same way that our skin heals when cut?

Once the group hit on this topic, it began free-associating. Could one apply regeneration to automobiles, robotic space probes (e.g. the Phoenix), and space shuttles? A biologist jumped in: how could we design a house that could repaint itself every other spring or replace the shingles on its roof after storms? Could we design roads and highways that would fill their own potholes?

The Questions Seemed Endless

Having begun the conversation wondering how to apply engineering concepts to biological systems, the group then asked whether understanding of biological systems could enhance the robustness of engineered systems. Engineers have long looked to biology for inspiration. The sleek and efficient body plan of the bottlenose dolphin has been exhaustively studied by submarine designers. Detailed study of the albatross wing aided aircraft manufacturers, and our newest super-computers attempt to incorporate our limited understanding of neural networks to increase processing speed.

Research proposals eluded the group but some felt that the discussions generated enough new ideas for a short perspective that could be published in a scientific journal. The perspective focuses on how engineered systems might learn from biological principles of regeneration to build more complex robust systems. Specifically, they are examining modular components at small scales. For example, using a limited number of building blocks (e.g. 21 amino acids) and using those blocks to build something new (e.g. heat shield on a spacecraft).

Task Group Summary 8
Ecological robustness: Is the biosphere sustainable?

CHALLENGE SUMMARY

Ecological systems provide services to humanity that support life as we know it. These "ecosystem services" include obvious and direct products such as food, fiber, fuel, and other goods that we extract from living beings; indirect benefits such as the mediation of climate and the sequestration of toxic substances; and, most subtly, aesthetic pleasure such as the provision of wilderness. There are those who believe that ecosystems have evolved regulatory mechanisms that will maintain resilience and robustness in the face of disturbances, or that technological advances will be able to compensate; yet, from an evolutionary point of view, such confidence cannot be justified. Ecosystems and the biosphere are complex adaptive systems, in which the selfish agendas of individual agents threaten the robustness of the whole. What is the adaptive potential of the biosphere to deal with climate change and other global and regional stressors, and how can we relate the robustness of macroscopic features to the microscopic dynamics of species? Can the modeling of coupled ecological and social systems provide the necessary feedbacks to prevent catastrophic shifts? Do we have an ethical obligation to preserve the regulatory and adaptive mechanisms of the world's ecosystems?

Key Questions

- Are there quantifiable and universal emergent properties of populations, communities, and ecosystems?

- Do scaling laws apply to ecological systems? Could the data ever be adequate to test hypotheses?
- Can we model the biosphere and its coupled processes with predictive capability? Can we prove that we cannot?
- Do food webs have common architectures? Signatures of self-organization?
- Co-evolution of biosphere and geosphere: How does life shape the non-living components of the planet?
- The biosphere and climate: Can the trajectory of this complex system be understood?
- How does the information encoded in genomes manifest itself in the biogeochemical properties of ecosystems (e.g., the ratios of elements in living matter)?
- How does cooperation among organisms emerge over the course of evolution?
- Are there common features of biochemical networks within the cell (systems biology), and networks of the ecosystems that are built from them?

Required Reading

Kirchner JW. The Gaia hypothesis: can it be tested?. *Reviews of Geophysics* 1989;27:223-235.

Levin SA. Ecosystems and the biosphere as complex adaptive systems. *Ecosystems* 1998;1:431436.

Levin SA, Barrett S, Aniyar S, Baumol W, Bliss C, Bolin B, Dasgupta P, Ehrlich P, Folke C, Gren IM, Holling CS, Jansson A, Jansson BO, Mäler KG, Martin D, Perrings C, Sheshinski E. Policy Forum: Resilience in natural and socioeconomic systems. Environment and Development Economics (1998), 3:221-262. Cambridge University Press. [Accessed online June 2, 2008: http://journals.cambridge.org/download.php?file=%2FEDE%2FEDE3_02%2FS1355770X98000126a.pdf&code=9c2b0dcbaeacab587fc0936b8a4b398c.]

TASK GROUP MEMBERS

- Luis Amaral, Northwestern University
- Elias Greenbaum, Oak Ridge National Laboratory
- Ross Hammond, The Brookings Institution
- Chris Klausmeier, Michigan State University
- Rob Knight, University of Colorado, Boulder

- Jennifer Martiny, University of California, Irvine
- Susanne Menden-Deuer, University of Rhode Island
- Caitlin O'Connell-Rodwell, Stanford University
- Joshua Weitz, Georgia Institute of Technology
- Alexandra Worden, Monterey Bay Aquarium Research Institute
- Lisa Song, MIT

TASK GROUP SUMMARY

By Lisa Song, Graduate Science Writing Student, MIT

The drowning polar bear of the Arctic is a vivid symbol of climate change's impact in shaping the life or death of a species. The threat facing the polar bear is representative of what happens when the footprint of human activities becomes so vast that if affects the quality of water, land, and air, and even the temperature of our planet. Today, climate change is just one of many causes pushing ecosystems into decline, but the polar bears are merely the most visible symbol of a deeper problem. At the 2008 National Academies Keck *Futures Initiative* Conference on Complex Systems, a group of scientists from interdisciplinary fields tackled the following question: is the biosphere sustainable? To answer the question, which includes concepts of ecological robustness, requires expertise on all the inner workings of an ecosystem, from microbes and plants to nutrient cycling and atmospheric chemistry.

At the moment, what is known about the ecosystem is vastly overshadowed by what is still unknown. Biosphere 2, the largest *in vivo* recreation of Earth's ecosystems, was built in the Arizona desert in the late 1980's. Eight humans spent two years within its walls during 1991-1993, but the experiment failed spectacularly when swarming ants took over the land and the residents required supplemental oxygen for survival. If we are to create a habitable Earth on Mars, or sustain the ecosystems we already have on Earth, we must first know where we stand today, and what is needed to create a sustainable biosphere.

The Group's Approach

The group began with the consensus that our current patterns of resource consumption and habitat destruction are not sustainable. As fisheries collapse, savannas turn to desert, and thousands of species hurtle towards extinction, we are degrading and, in some cases, irreversibly destroying the world's ecosystems. The human population is expected to reach around 9 billion by 2050, so the scope of our impact is likely to increase with time.

The group defined sustainability as the ability to use the resources of the biosphere without depletion in the long-term. To use a financial analogy, if the biosphere is an investment, how can we limit ourselves to using the interest rather than the principal? A major limitation to formulating solutions is the severe lack of understanding of many, if not most, ecosystem functions. Moreover, a lack of long-term observational data makes it difficult to discern trends from variation. Nonetheless, the depletion of resources and significant shifts in long-term and large-scale patterns in the biosphere is evident and undoubtedly caused by human intervention. Due to the urgency of the problem, solutions need to address the problems, without further exasperating them through unknown interactions and feedbacks.

Solutions

The group split its solution into five sections: modeling, data, experiments, interventions, and resources. The first three are aimed at understanding processes to guide human actions and policy. Interventions will flow from these results; resources will be needed to turn these plans into reality.

Modeling

In the documentary *An Inconvenient Truth,* Al Gore makes repeated use of the hockey-stick graph showing global temperature increases over time. Developed by scientists, models are powerful tools for policy-makers. The best models will present likely scenarios for the future and offer pragmatic paths that characterize different possible outcomes.

In order to understand the biosphere as a whole, it is necessary to develop models that can integrate processes across vast spatial and temporal scales. Currently, there is a lack of consistency across data sets. Different models use measurements and units that are incompatible with one another, and the general lack of connectivity limits the use of these models. The group proposed the creation of a World Biosphere Organization (WBO) to integrate and coordinate ecosystem research and sustainability efforts. The WBO would also promote the accessibility of data by requesting that ecology journals only publish articles from scientists who will share their research for free (either online or through print).

Modeling is an iterative process that changes with time, but it can be improved through competition and review. A good example is CASP (Critical Assessment of Techniques for Protein Structure Prediction), an online forum through which protein structures are proposed, tested, and validated or refuted. The WBO could use CASP as an example to promote the testing of ecological models against one another, in order to increase quality and to keep the models up-to-date.

Data

Ecologists are awash in streams of data that capture moments in time. Buoys in the ocean can monitor temperature, salinity, and pH; biologists can manually count the number of monarch butterflies in one area of the world, but integrated, time-series data on processes such as nutrient cycling, photosynthesis, and biogeochemical cycles are largely missing. Without information about rates, fluxes, and flows, it's impossible to gauge the connections in an ecosystem. We need a concerted, worldwide effort to create devices that monitor ecological processes.

In addition, data collection is unevenly distributed around the world. Detailed satellite data are often concentrated in developed countries, but ecosystems do not respect international boundaries. To extend the network of data-gathering, it would be helpful to promote the participation of citizen scientists, using cell phones as a vehicle for communicating data. Many developing countries are "leapfrogging" over traditional barriers to development by promoting cell phone use instead of landlines. The phones have computational power, so cell phone users who sign onto the program would receive a small compensation for their help in monitoring local ecosystems. The data would be stored in their cell phones and transferred through the Internet. Development of the data-transfer systems would be done through the WBO.

Having millions of citizen scientists around the world would allow for fast on-the-ground monitoring. If something unusual happens, ecologists can immediately focus their efforts on the issue in much the same way that the CDC rushes to the site of an epidemic.

Experiments

Halfway between the extravagance of Biosphere 2 and the simplicity of glass-blown EcoSpheres (miniature marine ecosystems sold as paperweights), mesocosms are ideal for ecosystem experimentation. A series of mesocosms, collectively called Biosphere Exploratory Experiments, could bridge the gap between lab work and *in vivo* experimentation. The mesocosms would cover a range of ecosystems scales; some would contain just dirt, microbes, and water, while others will have more complexity. Like mini experimental stations, the mesocosms would be used to discover connections, feedback loops, and connectivity within ecosystems.

It's important to note that most life forms are microbial, unknown, and, at present, unculturable. We have a lot to learn about microbial life, but by experimenting on multiple scales, we can track and model the influence of microbes in ecosystems.

Intervention

While the proposed WBO will coordinate research worldwide, a similar organization for the United States could operate productively on a national level. Unlike the EPA which is focused on regulations, the new group (which could be called the National Institute of the Environment) would concentrate on research alone.

Public outreach and education are essential for spreading the word about ecological threats. The food pyramid is an ideal example of outreach, and its design is simple, attractive, and easy to understand. More important, the pyramid offers consumers a way to visualize changes in their personal lives. The group proposed creating similar pyramids for energy, water, policy, economics, etc. based on a sustainability gradient.

The energy pyramid could look like this the one on the following page.

It's important to emphasize how different pyramids could interact. Meat is an essential part of the food pyramid, but cows produce large quantities of methane that contribute to global warming. The pyramids would allow the public to map the connections between ecosystems, human actions, and sustainability.

Ultimately, the group's goal is to promote a preventionist approach. Instead of cleaning up the mess after we destroy an ecosystem, the models and experiments will reveal projections of ecosystem futures, and what we

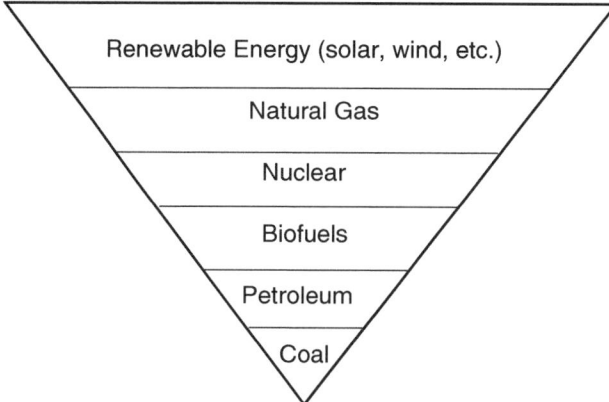

can do to steer the biosphere towards a path that will sustain humans for generations to come.

Resources

And finally, none of the plans would be possible without adequate funding. The annual US budget spent on environmental and ecological research is about $2 billion. A 1987 study estimated that the Global Biosphere Product was US $33 trillion, nearly double the 1987 Gross World Product of US $18 trillion. The numbers show a clear discrepancy between the funds allocated to R&D and the high cost of damaging the ecosystem. A mechanistic understanding of ecosystems functions is the first step towards sustainability; more funds must be appropriated in order to understand how the biosphere works, where it is headed, and what we can do to maintain its continued prosperity.

Task Group Summary 9
Can one control flow and transport in complex systems?

CHALLENGE SUMMARY

Transport in complex systems involves the flow of a quantity—information, power, mass, material, etc.—among the individual component elements. The nature of this transport depends both on the properties of the individual components and on the overall geometrical and topological structure of the system. In simple physical systems, we typically find either "ballistic" transport—that is, the distance travelled is proportional to time, as in the flight of a projectile—or "diffusive" transport—that is, the distance traveled is proportional to the square-root of time, as in Brownian motion describing the spread of a drop of dye in an unstirred liquid. In complex systems, as a consequence of inherent nonlinearities and complicated connectivity, transport becomes—no pun intended—much more complex. Two examples related to but distinct from standard diffusion will illustrate this complexity. Random walks in which the increments are distributed according to distributions with "fat tails" (instead of Gaussians) are known to produce "Lévy flights," in which interspersed with the small jumps typical of Brownian motion are long jumps that lead to "anomalous" (super-diffusive) transport. Such Lévy flights can occur in the spatial or temporal domain and are observed in analysis of data from earthquakes, finance, fluid flows, and animal foraging, among many other systems. When nonlinear effects—arising from predator-prey interactions or chemical reactions—are added to diffusion, the resulting "reaction-diffusion" equations can exhibit pattern-forming instabilities that can lead to "morphogenesis" or to wave-like transport with ballistic properties. A celebrated example is the model studied by Murray for the spread of a potential rabies epidemic in England. Murray showed that the underlying model (based on the Fisher equation) led to a narrow wave-front of contagion moving at a definite speed through the countryside.

When we consider complex systems involving networked structures, the problem becomes, technically, the study of transport on (arbitrary) graphs. Intuitively, it is clear that the nature of the network/graph—e.g., hub and spoke, long-range connections, random, etc.—will affect the transport dramatically.

Our attempt to understand the nature of transport in complex systems is in large part driven by the goal of controlling this transport. In some cases, we want to enhance transport: for instance, increasing the ability of the Internet to carry messages, enhancing traffic flow, increasing the rate of oil recovery or the efficiency of mixing and disseminating information in the case of a crisis, etc. In other cases, we want to inhibit transport—halting the spread of a virus or other contagion, preventing the collapse of an economic structure (the savings and loan sector, home mortgage sector), etc. *A priori*, we can imagine controls that work on the components (nodes) of the system—e.g., changing interest rates or leverage requirements, vaccinating individuals—as well as controls that work on the connections (links)—changing diffusion constants, limiting travel, severing links. Overall, the challenge is to use these various controls to manage transport in a complex system so as to optimize it for a desired outcome.

To define this challenge more precisely, different task groups might consider one or more of the following four "case studies" from four very different disciplines.

1. In economic systems, the recent sub-prime mortgage fiasco represented "transport" by catastrophic cascading collapse; might it have been avoided if, in addition to interest rates, the government controlled the "leverage" that firms could use? The infamous collapse in 1998 of Long Term Capital Management (LTCM), which failed due to margin calls, had its origin in an incorrect evaluation (by the world's experts!) of the actual risks involved in some of the investments; technically, their models failed to take account of the "fat tails" of the risk distribution. Can we create, before the fact, reliable models of risks in complex economic structures (or, what part of "derivatives" don't we understand)?

2. In oncology, we need to consider transport at both the molecular and organism levels. At the molecular level, cancer is usually a disease

caused by mutations in genes important for cellular regulation such as cell cycle, development, apoptosis, etc. Although undeniably a good start, this description of cancer fails to explain fully the progression from quiescent, non-cancerous, to fully malignant and eventually metastatic cells and the accumulation of multiple cancer mutations along the way. Macromolecules such as proteins and RNAs encoded by cellular genes interact with each other to form a molecular dynamic system of great complexity. The systems properties of such molecular "interactome" networks have remained largely unknown until recently, primarily due to the lack of empirical description. In the aftermath of the human genome sequencing project, systems biologists are developing concepts, tools, and resources to model interactome networks with the goal of modeling differences of systems properties between cancerous and non-cancerous cellular networks. The ultimate goal of this endeavor is the design of drugs that would be able to alter systems properties of cancer cells to either kill them specifically or dramatically slow their malignant progression. At the organism level, we need to consider both how the primary tumor "transports" its malignancy—basically, by rapid localized (diffusive?) growth and displacement of normal cells—and how secondary tumors are created by metastasizing cells transported through the body by lymph or blood networks (Lévy flights?). Can we develop (perhaps different) appropriate therapies that will be needed to attack these two different forms of transport?

3. In public health systems, the challenges are both highly visible and daunting. Preventing the spread of various epidemics—SARS, Avian flu—and limiting the damage of the AIDS pandemic are among the most important problems facing society today. It is important to recognize that air travel—quite literally, a Lévy flight—played a significant role in the initial spread of AIDS between San Francisco and New York and the later studies that showed (*post-hoc*) that the spread of the SARS epidemic could have been predicted by air travel patterns, suggesting that restricting such travel in times and from regions of high contagion might be, despite its Draconian nature, an appropriate policy. Would this really be a workable and effective policy?

4. Much of our key societal infrastructure exists in the form of networks—the electrical power grid and the Internet are two important examples. The celebrated Northeast electrical blackout of 1965 was thought to have provided a transformative lesson, but a very similar cascading failure occurred in 2003 and likely could occur again. What lessons should we have learned from these failures? How can we control the system so as to keep the

effects of power plant failures localized? Regarding the Internet, there are at least two key questions. First, the rapid spread of computer viruses with pandemic consequences is enabled by the Internet: can we develop a means of identifying these viruses as they travel and prevent them from attacking individual computers (i.e., severing the links)? Alternatively, the "monoculture" of operating systems renders the individual computers much more susceptible to viral attacks: can we design operating systems that are sufficiently individualized so as resist these attacks (i.e., modify the nodes). Second, recent studies have shown that the Internet itself is particularly vulnerable to attacks on its key hubs; how can we improve the systems to make it more resistant to these attacks? Clearly, both of these infrastructure "transport" issues overlap very strongly with the studies of robustness in other task groups.

Key Questions

- In addition to the questions already posed in the individual case studies above, are there other overarching questions that we should consider?
- To what extent do the transport mechanism given in the examples above exhaust those likely to be found in complex systems? What can we add to this "taxonomy of transport"?
- Are there any universal aspects of transport in complex systems?
- What instructive "case study" examples can we find from other disciplines?
- What is the optimal mix of controls on the nodes versus controls on the links? How does this vary across different complex systems?
- How should we proceed to develop strategies to enhance desired flows and to inhibit undesired flows?

Required Reading

For Lévy Flights

Solomon T, Weeks E, and Stanley H. Observations of anomalous diffusion and Lévy Flights in a two-dimensional rotating flow," *Phys Rev Lett* 71, 3975-3979. [Accessed online July 31, 2008: http://www.physics.emory.edu/~weeks/abs/nice95.html.]

Geisel T, Nierwetberg J, and Zacherl A. Accelerated diffusion in Josephson junctions and related chaotic systems. *Phys Rev Lett* 1985;54:616-620.

For the Study of the Spread of Rabies in England

Murray JD, Stanley EA, and Brown DL. On the spatial spread of rabies among foxes. *Proc Roy Soc (Lond)* 1986;B229:111-150.

Murray JD. Modeling the spread of rabies. *American Scientist* 1987;(May-June):280-284.

For Cancer

Huang S and Ingber DE. A non-genetic basis for cancer progression and metastasis: self organizing attractors in cell regulatory networks. *Breast Disease* 2007;26:27-54.

Kitano H. Cancer as a robust system: implications for anticancer therapy. *Nature Reviews Cancer* 2004;4(Mar):227-235.

For Public Health and Epidemics

Hufnagel L, Brockmann D, Geisel T. Forecast and control of epidemics in a globalized world. *Proc Natl Acad Sci USA* 2004;101:15124-15129.

Severe Acute Respiratory Syndrome (SARS) background. Wikipedia reference. [Accessed online August 13, 2008: http://en.wikipedia.org/wiki/SARS.]

For Economic "Collapsing Cascades"

Long-Term Capital Management. Wikipedia reference. [Accessed online August 13, 2008: http://en.wikipedia.org/wiki/LongTerm_Capital_Management.]

Report of the CRMPG III August 6, 2008 (Counterparty Risk Management Policy Group III). Containing systemic risk: the road to reform.[Accessed online August 13, 2008: http://www.crmpolicygroup.org.]

For Power Grid Failures

Kinney R, Crucitti P, Albert R, Latora V. Modeling cascading failures in the North American power grid. *Eur Phys J* 2005;B46:101-107.

Suggested Reading

Barabási AL. The day the lights went out; we're all on the grid together. *New York Times* 2008. Opinion. [Accessed online August 13, 2008: http://query.nytimes.com/gst/fullpage.html?res=950CE5D91430F935A2575BC0A9659C8B63&scp=9&sq=barabasi&st=cse.]

Goldberger AL. Non-linear dynamics for clinicians: chaos theory, fractals, and complexity at the bedside. *The Lancet* 1996;437:1312-1314.

Källén A, Acuri P, and Murray JD. A simple model for the spatial spread of rabies. *J Theor Bio* 1985;116:377-393.

Lengvel I and Epstein IR. A chemical approach to designing Turing patterns in reaction diffusion systems. *Proc Natl Acad Sci USA* 1992;89:3977-3979.
Levine H and Rappel WJ. Membrane bound Turing patterns. *Phys Rev E* 2005;72:061912.
Mandelbrot B. *The Fractal Geometry of Nature.* W.H. Freeman and Company 1982.
Murray JD and Seward WL. On the spatial spread of rabies among foxes with immunity. *J Theor Biol* 1992;156:327-348.
Turing AM. The chemical basis of morphogenesis. *Philosophical Transactions of the Royal Society B (London)* 1952;237:37-72.

Due to the popularity of this topic, two groups explored this subject. Please be sure to review the second write-up, which immediately follows this one.

TASK GROUP MEMBERS – GROUP A

- Marta Gonzalez, Northeastern University
- Challa S.S.R. Kumar, Louisiana State University
- Ying-Cheng-Lai, Arizona State University
- Shayan Mookherjea, University of California San Diego
- Frederick Moxley II, United States Military Academy
- Michael J. North, Argonne National Laboratory
- Juan Ocampo, Trajectory Asset Management
- Iraj Saniee, Bell Laboratories, Alcatel-Lucent
- Alessandro Vespignani, Indiana University
- Anne-Marie Corley, MIT

TASK GROUP SUMMARY – GROUP A

By Anne-Marie Corley, Graduate Science Writing Student, MIT

A group of scientists, representing many disciplines at the 2008 National Academies Keck *Futures Initiative* Conference in Irvine, California, was asked to consider this question: Can one control flow and transport in a complex system?

Understanding the nature of transport in complex systems is essential to controlling it. In some cases the goal is to enhance transport: for instance, increasing the ability of the Internet to carry messages, enhancing traffic flow, increasing the rate of oil recovery, or increasing the efficiency of disseminating information in a crisis. In other cases, the goal is to inhibit transport, for example halting the spread of a virus or preventing the collapse of an economic structure.

The first challenge in rallying diverse backgrounds to consider this question was to ask how the group might constrain the systems under examination. Would they analyze systems under duress (such as market crisis and power grid failure), or look at systems under normal conditions (healthy markets, smoothly functioning power grids, 'green-light' conditions for transportation systems)? In other words, at what point did they seek to control flow?

After considering a range of ideas such as detecting the next zoonotic disease to hop from animals to humans, locating networks of terrorists interacting in space and time, or controlling transportation flow to inhibit the spread of epidemics, the group decided to focus on two example application areas—namely the financial system's credit flow and the role of commuting patterns in the spread of epidemics. The group placed primary emphasis on the financial system's credit flow and used the spread of epidemics example as a check for logical clarity.

Once the example application areas were selected, the scientists developed a two-pronged approach, aiming first to *detect* an economic system in peril, and then to *control* it. This led to the identification of a *Detection Problem* and a *Control Problem*. The Detection Problem is 'what are the key observable features of the transition from normal uncorrelated (i.e., de-coherent) behavior to abnormal correlated (i.e., coherent) behavior?' The Control Problem is that 'given a system with amenable characteristics, are there parsimonious and indigenous mechanisms to "control" flow and transport and how do you chose among them?' Parsimonious here means minimum energy or cost. Indigenous here means a natural or normal part of the system.

The group felt that the concepts of coherence and decoherence were critical to solving both the Detection Problem and the Control Problem. Coherence occurs when the individual behavior of different actors in a system correlate. What it means is that individual actions converge from random, independent actions to dependent, 'matching' behavior. The classic economic example is a run on a bank. Think *It's a Wonderful Life*.

Many aspects of both human engineered and evolved systems depend on uncorrelated behaviors for proper functioning. For example, when behavior is comparatively random—decoherent, or uncorrelated—an economic system proceeds as normal. Some people put money into banks while others draw it out, but they do so independently. In this case decoherence is the norm, while correlated coherent behavior presents a problem. It can lead to the kind of convergence that brings down markets.

Obviously, real systems always exist in a state somewhere between complete coherence and decoherence. Systems run into trouble when the level of coherence and decoherence, or more generally the distributions of individual behaviors, diverge significantly from the original design assumptions or evolutionary conditions. It is also possible for coherence to be the normal healthy state and for decoherence to be a problem. However, as the bank run example suggests, excessive coherence in particular is a major threat to economic systems.

Detecting whether a system is about to head into a coherent phase is essential to influencing the outcome. The group hypothesized that there are indeed markers to indicate the onset of coherence. If they could then postulate a detection system, they might also imagine a way to intervene, turn the control valve, and keep the system from excessively cohering.

To address the Detection Problem the group asked, 'can we develop something analogous to an afferent (i.e., sensory) nervous systems for complex systems?' Financial system examples include: (1) number of web browser read accesses of official informational web sites, (2) patterns of search engine queries (e.g., Google searches), and (3) patterns of visits to relevant non-official web sites (e.g., number of web browser read accesses of Wikipedia entries). For epidemic spread in transportation systems examples include: (1) patterns found in mobile device usage (e.g., global positioning system or mapping data access on mobile phones), (2) geo-coded patterns of search engine queries (e.g., Google searches), and (3) geo-coded patterns of visits to relevant web sites (e.g., Wikipedia entry accesses). Along the way the group recognized that some organizations such as Google are actively working in these areas. The group concluded that these efforts should be cooperatively leveraged as much as possible.

To address the Control Problem the group hypothesized that the transition from decoherence to coherence is triggered by psychological contagion amplified by feedback loops. This creates the opportunity for early, low energy interventions and thus efficient control.

The group felt that modeling may be important. They postulated a conceptual model of the financial system that takes into account the economic, psychological, and social drivers of decision making. They also postulated that strong parallels hold for epidemic spread in transportation systems.

A critical research gap is real time psychological and behavioral data on human decision making that drives financial and transport feedback loops. In particular, what are the tipping points in the drive toward coherent behavior, when many people make the same move? How does accidental correlation factor in? Such information could aid in the creation of a model to predict or control the flow and transport of cash, emotion and ideas in an economic system.

TASK GROUP MEMBERS – GROUP B

- Lajos Balogh, Roswell Park Cancer Institute
- Peter Cummings, Vanderbilt University
- Martin Gruebele, University of Illinois
- Rigoberto Hernandez, Georgia Institute of Technology
- Maia Martcheva, University of Florida
- Saira Mian, Lawrence Berkeley National Laboratory
- Peter Sloot, University of Amsterdam
- Jeffrey Toretsky, Georgetown University
- Muhammad Zaman, University of Texas at Austin
- Brian Creech, University of Georgia

TASK GROUP SUMMARY – GROUP B

By Brian Creech, Graduate Science Writing Student, University of Georgia

Charged with the problem of how to control transport in complex systems, a group of scientists at the 2008 National Academies Keck *Futures Initiative* Conference on Complex Systems agreed that a control mechanism should be simple, impacting transport while also presenting the fewest negative effects on the health of that system. In Task Group (9B) two of the questions that arose are if there exists a single factor that affects the flow of a quantity across an entire system, and whether a system remains complex if it can be altered by a single variable.

The answer to the first question is yes: temperature. In several different systems, from the human body to deep sea ecologies, slight changes in temperature set off a complex series of reactions that change how things move across the system. The complexity of these systems amplifies small temperature changes across the entire network, resulting in the slowdown or cessation of entire processes in a series of chain reactions that affect the state of the system. A simple change in temperature can have damaging and irreversible consequences on the structure of the system. Taking the earth as an example of a complex system, global warming is a change in temperature across the entire system. A rise in temperature may increase the likelihood that non-native flora and fauna survive in polar regions. Temperature rises may lead to the development of new and sometimes catastrophic weather patterns, ocean levels rise, and animal and plant species die off due to a chain of events instigated by a subtle change in temperature.

One representation of a complex system is an (arbitrary) graph where transport of the quantity of interest—information, material, mass, energy—occurs over the nodes and links. Identifying optimal control points requires knowledge of the set of nodes and edges, the topology of the graph.

The group considered a "dynamic network model" of a complex system where nodes and edges appear and disappear over time. A node may itself represent a network at a lower level, in much the same way that network representations of organ systems can be viewed as subnetworks of a network representation of the organisms as a whole. When conditions change within organs and tissues, the conditions of transport across the entire body are changed.

The Task Group started to formulate dynamic network models of the spread of HIV/AIDS and cancer metastases. The human HIV/AIDS pandemic was viewed as a disease transported across a network where nodes correspond to individuals; cities are a series of dynamic nodes connected by airlines, with the disease being transported via the changing social/sexual connections among infected and non-infected individuals within the cities. Metastatic cancer was modeled as a network within the human body that uses the lymphatic system and the venous system to transport cancer from one organ to the next. The body's organs are themselves dynamic networks, and are subject to the same characteristics of a larger dynamic network. Both examples have important similarities, but their differences impact how diseases move across the networks.

Spread of HIV/AIDS Among Human Populations

The most notable difference between the spread of metastatic cancer and the spread of HIV/AIDS is the dynamic nature of the links in the HIV/AIDS network. Personal habits change; people quit using drugs or start using drugs, and old sexual connections fade while new ones are forged.

The spread of HIV exhibits the "birds of a feather flock together" phenomenon, where individuals within certain subcultures and socioeconomic groups are more likely to contract and spread HIV. Although effective methods for slowing down or inhibiting the spread of HIV/AIDS such as increasing the use contraceptives are known, personal habits and behaviors often confound such control methods. For example, lifestyle choices are highly individualized and notoriously hard to control, but education before the fact is less draconian than widespread quarantine afterwards. Thus, one good control mechanism is localized education campaigns that change the habits of enough individuals so that the disease becomes localized within smaller and smaller groups. Prevalent cultural attitudes, education level, and income all play a role in an individual's ability to be influenced by knowledge about how AIDS is spread, making education the most expensive and complicated means of control. Models that measure control need to account for these differences and reflect how effective certain types of education are among different groups of people, while also identifying the more mobile and connected groups that are more likely to spread the disease throughout the wider population.

Metastatic Cancer Spread

Cancer cells and cytokines, the molecules used in cellular communication, can move through the body via the lymphatic system, a system that closely resembles the network of train tracks within the United States. The key to finding a control mechanism to slow or stop the spread of cancer is to look for the most effective roadblocks for the system, and then find the heaviest traveled paths on which to place these roadblocks. This model looks at metastases—the spread of cancer cells to new parts of the body, for example malignant breast cancer cells moving to the bones—as a structural phenomenon. The body's immune system offers a means of implementing control. A model of metastatic cancer spread could be used to test how specific manipulations of the immune system might impede or encourage the growth of cancer as well as their impact on other parts of the body. In cancer patients, original tumors are often not the most dangerous; rather tumors that metastasize prove to be more deadly and less amenable to treatment. Elderly cancer patients tend to die of something other than cancer and show fewer signs of metastases.

It is important to remember that not much is known about the mechanisms that influence the spread of metastatic cancer. What is primarily needed are the data to help build a network architecture that matches observed patterns of the spread of metastatic cancer. Instruments or methods are needed to measure flow between the lymph nodes and organs. Endoscopic imaging techniques have been used to observe cancer cells in the gastrointestinal tract and may provide the necessary means of tracking the mechanism for metastatic spread.

Conclusions

The unique features of individual networks affect the patterns of flow within that network. In the case of metastatic cancer, there is evidence that cancer spreads from lymph node to lymph node. A fruitful territory to explore would be a model that mimics metastatic spread along the lymphatic system.

At this point, the key is to determine factors that impact flow across the network. By modeling metastatic cancer spread along the lymphatic system, it may be possible to learn how cancerous cells and cytokines move across the system, whether through a series of Lévy flights—random, long-range jumps into a new environment—or by diffusing across the system, signaling tumor growth within hospitable organs.

The question remains though, can a system still be complex if it is impacted by change in a single variable? A network's complexity amplifies change in the single variable across the entire system, but the consequences of that amplification can be damaging. One hopes that for metastases, simple roadblocks are found and that the solution is something as simple as changing the temperature of the body. As shown with the spread of HIV/AIDS, though, simple, wide ranging solutions, like draconian quarantines, can also limit the flow of benficient material across the entire system. Like education strategies geared towards local culture to prevent the spread of HIV/AIDS, what is needed to control metastatic spread is a mechanism that can be implemented locally, without affecting the healthy flow for the rest of the body. The relative simplicity or complexity of that solution remains to be seen.

Appendixes

Preconference Webcast Tutorials

September 24, 2008, 1:00 – 4:00 p.m. EDT
(10:00 a.m. – 1:00 p.m. PDT)

Spreading Processes and Complexity

Vittoria Colizza
Research Scientist
Complex Networks and Systems Laboratory
Institute for Scientific Interchange (ISI Foundation)

Emergence and Collective Phenomena in Equilibrium and Nonequilibrium Systems

Nigel Goldenfield
Swanlund Endowed Chair, Professor
Department of Physics, Institute for Genomic Biology
University of Illinois at Urbana–Champaign

Social Networks 101

David M.J. Lazer
Associate Professor of Public Policy
Director of the Program on Networked Governance
Harvard's Kennedy School

**September 25, 2008, 1:00 – 5:30 p.m EDT
(10:00 a.m. – 2:30 p.m. PDT)**

Scaling and Fractals

Shlomo Havlin
Professor
Department of Physics
Bar-Ilan University, Ramat-Gan, Israel

Robustness in Complex Systems

James B. Bassingthwaighte
Professor of Bioengineering and Radiology
Department of Bioengineering
University of Washington

Neurobiology

Charles F. Stevens
Professor and Vincent J. Coates Chair in Molecular Neurobiology
Molecular Neurobiology Laboratory
Salk Institute

Non-Linear Science 101

David K. Campbell
University Provost
Professor of Electrical Engineering and Physics
Boston University

**September 26, 2008, 1:00 – 3:00 p.m. EDT
(11:00 a.m. – 1:00 p.m. PDT)**

Robustness in Complex Systems/Network Threats

Andreas Wagner
Professor
Department of Biochemistry
University of Zurich

Networks and Connectedness

Alessandro Vespignani
Professor
Department of Informatics, Physics, and Statistics
Indiana University

Bonus Presentation: Available on CD-Rom Only

A History of the Concept of Creativity

Richard N. Foster
Managing Partner
Millbrook Management Group, LLC
Board Member, W.M. Keck Foundation

Agenda

Thursday, November 13, 2008

7:15 and 7:45 a.m.   Bus Pickup (From the Marriott Newport Beach to the Beckman Center)

7:30 a.m.   Registration (Beckman Center/Outside Auditorium)

7:30 – 8:30 a.m.   Breakfast (Beckman Center/Dining Room)

8:30 – 8:45 a.m.   **Welcome and Opening Remarks**
Dr. Harvey V. Fineberg, IOM President
H. Eugene Stanley, Steering Committee Chair
Richard N. Foster, W.M. Keck Foundation Board Member (video presentation) (Beckman Center/Auditorium)

8:45 – 9:45 a.m.   **Keynote Address**
"The Architecture of Complexity: From the Topology of the WWW to the Structuring of the Cell"
Albert-László Barabási, Center for Complex Network Research, Northeastern University; Department of Medicine, Harvard University (Auditorium)

9:45 – 10:15 a.m.   Break (Atrium)

10:15 – 10:30 a.m.   **Task Group and Grant Program Overview**
(H. Eugene Stanley) (Auditorium)

10:30 – 10:35 a.m.	**Getting the Most Out of Task Group Discussions** Philip LeDuc, Associate Professor, Mechanical and Biomedical Engineering, and Biological Sciences, Carnegie Mellon University (Auditorium)
10:35 a.m. – 12:30 p.m.	**Panel Discussion** (Auditorium) *Moderator* • H. Eugene Stanley, University Professor, Professor of Physics, Professor of Chemistry, Professor of Biomedical Engineering, Professor of Physiology (School of Medicine), and Director, Center for Polymer Studies, Boston University (Auditorium) *Panelists* • Albert-László Barabási, Center for Complex Network Research, Northeastern University; Department of Medicine, Harvard University • James B. Bassingthwaighte, Professor of Bioengineering and Radiology, University of Washington • David K. Campbell, Professor of Physics and Electrical Engineering and Provost, Boston University • Jennifer A. Dunne, Research Fellow, Co-Director, Santa Fe Institute, Pacific Ecoinformatics and Computational Ecology Lab • Nigel Goldenfeld, Professor, Department of Physics, University of Illinois at Urbana-Champaign • David M.J. Lazer, Associate Professor, Harvard Kennedy School, Harvard University • M. Elisabeth Paté-Cornell, Burt and Deedee McMurtry Professor and Chair, Department of Management Science and Engineering, Stanford University • Herbert Sauro, Associate Professor, Department of Bioengineering, University of Washington • Charles F. Stevens, Professor, Molecular Neurobiology Laboratory, Salk Institute • Alessandro Vespignani, Professor of Informatics and Cognitive Science, Adjunct Professor, Physics and Statistics, Indiana University
12:30 – 1:15 p.m.	Lunch (Dining Room) Setup for Poster Sessions 1 and 2 (Hallways A and B)
1:15 – 2:00 p.m.	Poster Session 1 (Hallway A) (see "Posters" tab in binder)
2:00 – 5:30 p.m.	Task Group Session 1 Various Meeting Rooms (see page 4 of this tab)

3:30 – 4:00 p.m.	Break (Atrium and Second Floor Hallway)
5:30 – 7:00 p.m.	Reception (Beckman Center/Fountain Courtyard)
5:45 – 6:30 p.m.	Poster Session 2 (Hallway B)
7:00 – 9:00 p.m.	Communication Awards Presentation and Dinner (Atrium)
9:00 p.m.	Bus Pickup (From Beckman Center to Marriott Newport Beach)
9:00 – 11:00 p.m.	Informal Discussions/Hospitality Room (optional) (Marriott Newport Beach/Sunset Room)

Friday, November 14, 2008

7:00 and 7:30 a.m.	Bus Pickup (From the Marriott Newport Beach to the Beckman Center)
7:15 – 8:00 a.m.	Breakfast (Beckman Center/Dining Room)
8:00 – 10:00 a.m.	Task Group Session 2 (Various Meeting Rooms) (see "Task Groups" tab in binder)
10:00 – 10:30 a.m.	Break (Atrium and Second Floor Hallway)
10:30 – noon	Task Group Reports (5-6 minutes per group) (Auditorium)
Noon – 1:30 p.m.	Lunch Dining Room
12:45 – 1:30 p.m.	Related Task Group Discussions (Groups 3A-3B, 6A-6B, 9A-9B) Groups 3A-3B **(Executive Dining Room (First Floor))** Groups 6A-6B **(Newport Room (First Floor))** Groups 9A-9B **(Huntington Room (First Floor))**
1:30 – 5:00 p.m.	Task Group Session 3 (Same Meeting Room as Sessions 1 and 2)
3:00 – 3:30 p.m.	Break (Atrium and Second Floor Hallway)
5:00 p.m.	Task Group representatives to drop off presentation at information/registration desk or upload to FTP site prior to 7:00 a.m. Saturday morning (Atrium)

5:00 – 6:30 p.m.	Reception (Beckman Center/Fountain Courtyard)
6:30 – 8:00 p.m.	Dinner (Beckman Center/Atrium)
8:00 – 8:30 p.m.	**Dinner Speaker** Murray Gell-Mann, Distinguished Fellow, Santa Fe Institute (Atrium)
8:30 p.m.	Bus Pickup (From Beckman Center to Marriott Newport Beach)
9:00 – 11:00 p.m.	Informal Discussions/Hospitality Room (optional) (Marriott Newport Beach/Sunset Room)

Saturday, November 15, 2008

7:00 and 7:30 a.m.	Bus Pickup (From the Marriott Newport Beach to the Beckman Center)
7:15 – 8:00 a.m.	Breakfast (Beckman Center/Dining Room)
7:15 a.m.	Stop by registration/information desk to arrange for taxi service if shuttle bus service at noon and 1:30 p.m. does not work with schedule (Beckman Center/Atrium/Registration and Information Desk)
8:00 – 9:30 a.m.	Task Group Reports (10-12 minutes per group) (Auditorium)
9:30 -10:00 a.m.	Break (Atrium)
10:00 – 11:00 a.m.	Task Group Reports (continued) (Auditorium)
11:00 – noon	Q&A Across All Task Groups (Auditorium)
Noon – 1:30 p.m.	Lunch (optional) (Dining Room)
Noon and 1:30 p.m.	Buses Depart for Hotel and Airport (Buses depart Beckman Center for Marriott Newport Beach and John Wayne (SNA) Airport)

Participant List

Luis Amaral
Associate Professor
Chemical and Biological
 Engineering
Northwestern University

Ananth Annapragada
Associate Professor
Health Information Sciences
University of Texas Houston

Craig Atwood
Associate Professor
Medicine
UW-Madison

Lajos Balogh
Professor of Oncology and
 Director of Nanotechnology
 Research
Radiation Medicine
Roswell Park Cancer Institute

Albert-László Barabási
Distinguished University Professor
Center for Complex Network
 Research,
Department of Physics
Northeastern University
Department of Medicine, Harvard
 University

Noah Barron
Reporter/Annenberg Graduate
University of Southern California

James B. Bassingthwaighte
Professor of Bioengineering and
 Radiology
Department of Bioengineering
University of Washington

Amy Bauer
Postdoctoral Research Associate
Theoretical Division
Los Alamos National Laboratory

John M. Beggs
Assistant Professor
Physics
Indiana University

Kirstie Bellman
Principal Director
Aerospace Integration Science
    Center
The Aerospace Corporation

Sally Blower
Professor
Semel Institute for Neuroscience &
    Human Behavior
David Geffen School of Medicine
University of California, Los
    Angeles

Stephen J. Bonasera
Assistant Professor
Division of Geriatrics, Department
    of Medicine
University of California, San
    Francisco

Edward Boyden
Benesse Career Development
    Professor
Media Lab, Department of Brain
    and Cognitive Sciences,
    Department of Biological
    Engineering
MIT

Alan Boyle
Science Editor
MSNBC.com

Cassandra Brooks
Graduate Science Writing Student
University of California, Santa
    Cruz

Lizzie Buchen
Graduate Science Writing Student
University of California, Santa
    Cruz

Timothy G. Buchman PhD, MD,
    FACS, FCCM
Edison Professor of Surgery,
    Professor of Anesthesiology
    and of Medicine
Department of Surgery
Washington University in Saint
    Louis

George Butler
President, Director, Producer
White Mountain Films

David K. Campbell
University Provost, Professor
Department of Physics and
    Electrical Engineering
Boston University

Tansu Celikel
Assistant Professor
Neurobiology
University of Southern California

Kee Chan
Assistant Professor
Health Sciences
Boston University

Fahmida N. Chowdhury
Program Director
Cross-Directorate Activities, Social, Behavioral, and Economic Sciences
National Science Foundation

Noshir S. Contractor
Jane S. & William J. White Professor of Behavioral Sciences
School of Engineering, School of Communication, and the Kellogg School of Management
Northwestern University

Jeffrey Cooper
Chief Innovation Officer
Intelligence, Security and Technology Group
SAIC

Anne-Marie Corley
Graduate Science Writing Student
MIT

Jennifer Couch
Chief
Structural Biology & Molecular Applications Branch
National Institutes of Health

Brian Creech
Graduate Scince Writing Student
University of Georgia

James Crutchfield
Professor of Physics,
Director, Complexity Sciences Center
Complexity Sciences Center and Physics Department
University of California at Davis

Barbara J. Culliton
President
The Culliton Group

Eugenio Culurciello
Assistant Professor
Electrical Engineering
Yale University

Peter Cummings
John Robert Hall Professor
Department of Chemical Engineering
Vanderbilt University

Ana Diez Roux
Professor
Epidemiology
University of Michigan

Ramanand Dixit
Assistant Professor
Biology
Washington University in St. Louis

Raissa D'Souza
Associate Professor
Mechanical and Aeronautical Engineering and Complexity Sciences Center
University of California, Davis

Jennifer Dunne
Research Fellow, Co-Director
Santa Fe Institute, Pacific
    Ecoinformatics and
    Computational Ecology Lab

Tuan Duong
Senior Researcher
Bio-Inspired Technologies and
    Systems Group
JPL/CIT

Rhonda Dzakpasu
Assistant Professor
Physics
Georgetown University

Nathan Eagle
Research Scientist
MIT/Santa Fe Institute

Nick Ellis
Professor
Psychology
University of Michigan

Veit Elser
Professor
Physics
Cornell University

Thierry Emonet
Assistant Professor
Molecular, Cellular, and
    Developmental Biology
    Department
Yale University

Joshua Epstein
Senior Fellow & Director
Center on Social and Economic
    Dynamics
The Brookings Institution

Doyne Farmer
Professor
Santa Fe Institute

James Ferrell
Professor
Chemical and Systems Biology
Stanford University

Susan Fitzpatrick
Vice President
James S. McDonnell Foundation

Harvey V. Fineberg
President
Institute of Medicine

Daniel Fletcher
Associate Professor
Bioengineering
University of California, Berkeley

Richard N. Foster
Managing Partner
Millbrook Management Group,
    LLC
Board Member, W.M. Keck
    Foundation

Stuart Fox
Graduate Science Writing Student
New York University

PARTICIPANTS

Ken Fulton
Executive Director
National Academy of Sciences

James Gardner
Author and Attorney
Gardner & Gardner,
    Attorneys, PC

Murray Gell-Mann
Distinguished Fellow
Santa Fe Institute

James Glazier
Director, Biocomplexity Institute
Department of Physics
Indiana University

Nigel Goldenfeld
Professor
Department of Physics
University of Illinois at
    Urbana-Champaign

Marta Gonzalez, PhD
Postdoctoral Research Associate
Northeastern University

Rebecca Goolsby
Program Officer
Human and Bioengineering
    Systems
Office of Naval Research

Elias Greenbaum
Corporate Fellow
Chemical Sciences Division
Oak Ridge National Laboratory

Robert A. Greenes
Ira A. Fulton Chair and Professor
Biomedical Informatics
Arizona State University

Tony H. Grubesic
Assistant Professor
Geography
Indiana University

Martin Gruebele
Professor
Chemistry, Physics, Biophysics and
    Computational Biology
University of Illinois

Natali Gulbahce
Post-Doctoral Research Fellow
Center for Cancer Systems Biology
Dana Farber Cancer Institute;
Center for Complex Networks
    Research
Northeastern University

Ross Hammond
Fellow
Economic Studies
The Brookings Institution

John Hartman
Assistant Professor
Genetics
University of Alabama at
    Birmingham

Anne Heberger
Senior Evaluation Associate
National Academies Keck *Futures*
    Initiative
The National Academies

Monica Heger
Graduate Science Writing Student
New York University

Sarah Heilshorn
Assistant Professor
Materials Science and Engineering
Stanford University

Jessica Hellmann
Assistant Professor
Biological Sciences
University of Notre Dame

Rigoberto Hernandez
Associate Professor
Chemistry & Biochemistry
Georgia Institute of Technology

Amy Herr
Assistant Professor
Bioengineering
University of California, Berkeley

Chen Hou
Postdoctoral Fellow
Santa Fe Institute

Paul Humphreys
Professor
Philosophy
University of Virginia

Alan J. Hurd
Director
Lujan Neutron Scattering Center
    at LANSCE
Los Alamos National Laboratory

Ted Jackson
Staff Photographer
The Times-Picayune

Barbara Jasny
Deputy Editor
Science/AAAS

George A. Kaplan
Thomas Francis Collegiate
    Professor of Public Health
Center for Social Epidemiology
    and Population Health/Dept.
    of Epidemiology
University of Michigan

Christopher Kello
Associate Professor
Cognitive Science
University of California, Merced

Don Kennedy
President Emeritus and Bing
    Professor of Environmental
    Studies
Institute for International Studies
Stanford University
Editor in Chief Emeritus, *Science*

Chris Klausmeier
Assistant Professor
W.M. Kellog Biological Station
Michigan State University

Rob Knight
Assistant Professor
Chemistry & Biochemistry
University of Colorado, Boulder

Stephen J. Kron
Associate Professor
Molecular Genetics and Cell
    Biology
The University of Chicago

Challa S.S.R. Kumar
Director
Nanofabrication & Nanomaterials
Center for Advanced
    Microstructures and Devices
Louisiana State University

Pradeep Kumar
Fellow
Center for Studies in Physics and
    Biology
Rockefeller University

Ying-Cheng Lai
Professor
Electrical Engineering
Arizona State University

Paul Laibinis
Professor
Chemical and Biomolecular
    Engineering
Vanderbilt University

Arthur Lander
Professor
Developmental and Cell Biology
University of California, Irvine

James S. Langer
Professor
Department of Physics
University of California, Santa
    Barbara

Daniel Lathrop
Professor and Director
Physics and Geology
University of Maryland

David Lazer
Associate Professor
Harvard Kennedy School
Harvard University

Philip LeDuc
Associate Professor
Mechanical and Biomedical
    Engineering, and Biological
    Sciences
Carnegie Mellon University

Jennifer Lauren Lee
Graduate Science Writing Student
University of Southern California

Hadley Leggett
Graduate Science Writing Student
University of California, Santa
    Cruz

Rachel Lesinski
Program Associate
National Academies Keck *Futures
    Initiative*
The National Academies

Brian Litt
Associate Professor
Neurology and Bioengineering
University of Pennsylvania

Wolfgang Losert
Associate Professor
Physics
University of Maryland

Stefan Maas
Assistant Professor of Molecular
  Biology
Biological Sciences
Lehigh University

Patricia L. Mabry
Behavioral Scientist/Health Science
  Administrator
Office of Behavioral and Social
  Sciences Research (OBSSR)
National Institutes of Health

Robert Marshall
Outdoor Recreation Editor
The Times-Picayune

Maia Martcheva
Associate Professor
Department of Mathematics
University of Florida

Jennifer Martiny
Associate Professor
Ecology and Evolutionary Biology
University of California, Irvine

Davis Masten
Chair
Presidents' Circle
The National Academies

Grant McCall
Assistant Professor
Anthropology
Tulane University

Susanne Menden-Deuer
Assistant Professor
Oceanography
University of Rhode Island

Saira Mian
Staff Scientist
Life Sciences Division
Lawrence Berkeley National
  Laboratory

John Miller
Professor of Economics and
Social Science, Research Professor
Social and Decision Sciences
Carnegie Mellon University/Santa
  Fe Institute

Shayan Mookherjea
Assistant Professor
Electrical and Computer
  Engineering
University of California San Diego

Cristopher Moore
Associate Professor
Computer Science Department,
  Department of Physics and
  Astronomy
University of New Mexico

Frederick Moxley
Director of Research for Network
  Science
United States Military Academy

Christopher Myers
Senior Research Associate
Computational Biology Service
  Unit / Biotechnology Center
Cornell University

Hamid Najib
Personal Computer and Program
  Support Specialist
The National Academies

Roger Narayan
Associate Professor
Joint Department of Biomedical
  Engineering
University of North Carolina at
  Chapel Hill

Michael J. North
Deputy Director
Center for Complex Adaptive
  Agent Systems Simulation
Argonne National Laboratory

Juan Ocampo
CEO and Founder
Trajectory Asset Management

Caitlin O'Connell-Rodwell
Assistant Professor (Consulting)
Otolaryngology, Head & Neck
  Surgery
Stanford University

Maureen O'Leary
Director, Public Information
Office of News & Public
  Information
The National Academies

Panos Papadopoulos
Professor
Mechanical Engineering
University of California, Berkeley

Michael Leo Parchman
Associate Professor
University of Texas Health Science
  Center
South Texas Veterans Heatlh Care
  System

M. Elisabeth Paté-Cornell
Burt and Deedee McMurtry
  Professor and Chair
Department of Management
  Science and Engineering
Stanford University

Vimla L. Patel
Professor and Vice Chair,
  Biomedical Informatics
Director, Center for Decision
  Making and Cognition
Arizona State University

Joshua Plotkin
Assistant Professor
Department of Biology
University of Pennsylvania

Alan L. Porter
Evaluation Consultant
Technology Policy & Assessment Center
Georgia Institute of Technology

Richard Puddy
Behavioral Scientist
Division of Violence Prevention, Program Implementation and Dissemination
Centers for Disease Control and Prevention

Casey Rentz
Graduate Science Writing Student
University of Southern California

Aristides Requicha
Gordon Marshall Chair in Engineering
Computer Science and Electrical Engineering
University of Southern California

Luis Rocha
Associate Professor
Informatics
Indiana University

David Roessner
Evaluation Consultant
National Academies Keck *Futures Initiative*
The National Academies

Stephen Ryan
President
Doheny Eye Institute
Board member, W.M. Keck Foundation

Iraj Saniee
Director
Mathematics of Networks and Complex Systems
Bell Laboratories, Alcatel-Lucent

Jordan Sarver
Graduate Science Writing Student
University of Georgia

Herbert Sauro
Associate Professor
Department of Bioengineering
University of Washington

Michael A. Savageau
Distinguished Professor
Department of Biomedical Engineering and
Microbiology Graduate Group
University of California, Davis

Suzanne Scarlata
Professor
Physiology & Biophysics
Stony Brook University

Steven Schiff
Director, Center for Neural Engineering
Engineering Science and Mechanics, Neurosurgery, Physics
Penn State University

Mark Schleifstein
Environment and Conservation
   Reporter
The Times-Picayune

Eric Schwartz
Graduate Science Writing Student
Science Journalism Program
Boston University

Caterina Scoglio
Associate Professor
Electrical and Computer
   Engineering
Kansas State University

William Skane
Executive Director
Office of News and Public
   Information
The National Academies

Peter Sloot
Professor
Computational Sciences
University of Amsterdam

Lisa Song
Graduate Science Writing Student
MIT

Olaf Sporns
Professor, Associate Chair
Department of Psychological and
   Brain Sciences
Indiana University

Dagmar Sternad
Professor
Department of Biology, Electrical
   and Computer Engineering
   and Physics
Northeastern University

H. Eugene Stanley
University Professor; Professor
   of Physics; Professor of
   Chemistry; Professor of
   Biomedical Engineering;
   Professor of Physiology
   (School of Medicine);
   Director, Center for Polymer
   Studies
Department of Physics
Boston University

Charles F. Stevens
Professor and Vincent J.
   Coates Chair in Molecular
   Neurobiology
Molecular Neurobiology
   Laboratory
Salk Institute

Gustavo Stolovitzky
Manager
Functional Genomics and Systems
   Biology
IBM Research

Kimberly Suda-Blake
Program Director
National Academies Keck *Futures*
   Initiative
The National Academies

Dan Swenson
Staff Artist
The Times-Picayune

Mercedes Talley
Program Director
W.M. Keck Foundation

Jeffrey Toretsky
Associate Professor
Oncology
Georgetown University

Andreas Trabesinger
Senior Editor
Nature Physics

Jessika Trancik
Postdoctoral Fellow
Santa Fe Institute

Shyni Varghese
Assistant Professor
Bioengineering
University of California, San Diego

Alejandra C. Ventura
Research Fellow
Internal Medicine, Division of
    Hematology/Oncology and
    Comprehensive Cancer
University of Michigan

Alessandro Vespignani
Professor
Department of Informatics,
    Department of Physics,
    Department of Statistics
Indiana University

Andreas Wagner
Professor
Department of Biochemistry
University of Zurich

Joseph Wang
Associate Professor
Aerospace and Ocean Engineering
Virginia Polytechnic Institute and
    State University

Qian Wang
Associate Professor
Chemistry and Biochemistry
University of South Carolina

Douglas Weibel
Assistant Professor
Biochemistry
University of Wisconsin, Madison

Leor Weinberger
Assistant Professor, Pew Scholar
Chemistry & Biochemistry
University of California, San Diego

Joshua Weitz
Assistant Professor
School of Biology
Georgia Institute of Technology

John P. Wikswo
Gordon A. Cain University
    Professor
Vanderbilt Institute for Integrative
    Biosystems Research and
    Education
Vanderbilt University

Alexandra Worden
Scientist/Assistant Professor
Monterey Bay Aquarium Research Institute

Lani Wu
Assistant Professor
Pharmacology & Systems Biology
University of Texas Southwestern

Larry Yaeger
Professor
Informatics
Indiana University

Muhammad H. Zaman
Assistant Professor
Department of Biomedical Engineering and Institute of Theoretical Chemistry
The University of Texas at Austin

Mingjun Zhang
Associate Professor
Biomedical Program Engineering, Department of Mechanical, Aerospace and Biomedical Engineering
The University of Tennessee, Knoxville